GeoPlanet: Earth and Planetary Sciences

More information about this series at http://www.springer.com/series/8821

Tymon Zielinski · Iwona Sagan
Waldemar Surosz
Editors

Interdisciplinary Approaches for Sustainable Development Goals

Economic Growth, Social Inclusion and Environmental Protection

Editors
Tymon Zielinski
Institute of Oceanology
Polish Academy of Sciences
Sopot
Poland

Waldemar Surosz
University of Gdansk
Gdynia
Poland

Iwona Sagan
University of Gdansk
Gdansk
Poland

The GeoPlanet: Earth and Planetary Sciences Book Series is in part a continuation of Monographic Volumes of Publications of the Institute of Geophysics, Polish Academy of Sciences, the journal published since 1962 (http://pub.igf.edu.pl/index.php).

ISSN 2190-5193 ISSN 2190-5207 (electronic)
GeoPlanet: Earth and Planetary Sciences
ISBN 978-3-319-89105-7 ISBN 978-3-319-71788-3 (eBook)
https://doi.org/10.1007/978-3-319-71788-3

Printed on acid-free paper

This Springer imprint is published by Springer Nature
The registered company is Springer International Publishing AG
The registered company address is: Gewerbestrasse 11, 6330 Cham, Switzerland

Series Editors

Managing Editor

Anna Dziembowska
Institute of Geophysics, Polish Academy of Sciences

Advisory Board

Kon-Kee Liu
Institute of Hydrological and Oceanic Sciences
National Central University Jhongli
Jhongli, Taiwan

Teresa Madeyska
Research Centre in Warsaw
Institute of Geological Sciences
Warszawa, Poland

Stanisław Massel
Institute of Oceanology
Polish Academy of Sciences
Sopot, Poland

Antonio Meloni
Instituto Nazionale di Geofisica
Rome, Italy

Evangelos Papathanassiou
Hellenic Centre for Marine Research
Anavissos, Greece

Kaja Pietsch
AGH University of Science and Technology
Kraków, Poland

Dušan Plašienka
Prírodovedecká fakulta, UK
Univerzita Komenského
Bratislava, Slovakia

Barbara Popielawska
Space Research Centre
Polish Academy of Sciences
Warszawa, Poland

Tilman Spohn
Deutsches Zentrum für Luftund Raumfahrt in der Helmholtz Gemeinschaft
Institut für Planetenforschung
Berlin, Germany

Krzysztof Stasiewicz
Swedish Institute of Space Physics
Uppsala, Sweden

Ewa Szuszkiewicz
Department of Astronomy and Astrophysics
University of Szczecin
Szczecin, Poland

Roman Teisseyre
Department of Theoretical Geophysics
Institute of Geophysics
Polish Academy of Sciences
Warszawa, Poland

Jacek Tronczynski
Laboratory of Biogeochemistry of Organic Contaminants
IFREMER DCN_BE
Nantes, France

Steve Wallis
School of the Built Environment
Heriot-Watt University
Riccarton, Edinburgh
Scotland, UK

Wacław M. Zuberek
Department of Applied Geology
University of Silesia
Sosnowiec, Poland

Piotr Życki
Nicolaus Copernicus Astronomical Centre
Polish Academy of Sciences
Warszawa, Poland

Preface

The impacts of climate change are already being felt on every continent, and according to the IPCC, the world's greenhouse gas emissions are continuing to increase. On January 1, 2016, the 17 Sustainable Development Goals (SDGs) of the 2030 Agenda for Sustainable Development officially came into force. There are 12 of the 17 SDGs that involve taking action on climate change. For sustainable development to be achieved, it is crucial to harmonize three interconnected core elements: economic growth, social inclusion, and environmental protection.

In this book, we provide scientific basis for a number of modern approaches and state-of-the-art methods of monitoring both the environment and social behavior and human expectations toward the protection of environment. The authors of the papers discuss these issues from different research perspectives, from physics to social sciences, and they point out the challenges and future scenarios based on the results of scientific activities.

Sopot, Poland
Gdansk, Poland
Gdynia, Poland
September 2017

Tymon Zielinski
Iwona Sagan
Waldemar Surosz

Contents

Introduction

Iwona Sagan, Waldemar Surosz and Tymon Zielinski

Climate change processes are affecting every region on every continent of our planet and thus disrupting all economies and influencing lives, on a daily basis. To make the situation worse, these changes will have a significant impact on humans and their communities as well as the entire natural environment for the next decades (IPCC 2013; Pelling et al. 2015).

Due to climatic changes we are experiencing changes, which have significant impact on every aspect of our environment with variations in weather patterns, rising sea level, and increasingly more frequent and more extreme weather events. The greenhouse gas emissions, which are a key factor in driving climate change, are at very high levels, highest in history, and unfortunately, continue to rise. The climate projections show that if nothing changes in terms of these emissions, the world's average surface temperature will be increasing over the 21st century and may exceed 3 °C over the next 100 years. This warming up will not be uniform, some regions of the world are likely to warm even more (IPCC 2013).

It is an undisputable fact that the climate change is caused by human activities and poses a major threat to the way we live and the future of our planet. In order to face these changes and make an attempt to create a sustainable world for everyone, humans need to address the climate change in the most comprehensive way (IPCC 2013; Ruwa et al. 2017a, b).

One of the most significant steps, and the real foundation to improving people's lives and fulfill the assumptions of the sustainable development is to provide quality education to everyone, independent of region, sex and status. In recent years, significant progress has been made towards increasing access to education at all levels, and as a result basic literacy skills have improved tremendously. This progress

T. Zielinski (✉)
Institute of Oceanology Polish Academy of Sciences, Sopot, Poland
e-mail: tymon@iopan.gda.pl

I. Sagan · W. Surosz
University of Gdansk, Gdansk, Poland

T. Zielinski et al. (eds.), *Interdisciplinary Approaches for Sustainable Development Goals*, GeoPlanet: Earth and Planetary Sciences,
https://doi.org/10.1007/978-3-319-71788-3_1

however, is far from satisfactory and even greater efforts must be made in order to make even greater leap to secure access to universal education by all people (UN; Fazey et al. 2016).

It is everyone's inalienable right to have access to clean, potable water. Humans strive to live in such world, and we know now that there are sufficient water resources on earth to achieve this. However, such common problems as water scarcity, its poor quality, often resulting in poor hygiene as well as low food security, have a variety of impacts on people's choices and educational opportunities across the world. (UN; Denton et al. 2014; Pelling et al. 2015).

Another environmental problem involves deforestation and desertification, both caused by human activities and climate change. These processes are of a key importance and are among the major challenges to obtain sustainable development.

Wild forests cover some 30% of the planet's surface. Not only do they serve as food supplies and shelters, forests play a major role in combating climate change effects, throughout the protection of biodiversity and providing homes to many indigenous populations. Nevertheless, it is as much as 13 million hectares of forests that are being destroyed every year and the further degradation of drylands has led to the desertification of 3.6 billion hectares (UN). People are getting increasingly aware of the importance of making every effort to manage forests and combat desertification.

Climate change is a global issue, no borders stop the changes, since e.g. adverse emissions somewhere will sooner or later affect people somewhere else. Therefore, climate change requires solutions that need to be addressed and hence coordinated at the international level and international cooperation is necessary to fight the adverse impacts of the global changes.

The world had turned in 2015, when representatives of most world countries adopted the 2030 Agenda for Sustainable Development and its 17 Sustainable Development Goals (SDGs). On 1 January 2016, the 17 Sustainable Development Goals of the 2030 Agenda for Sustainable Development came officially into force. Twelve out of the 17 SDGs involve taking action on climate change.

Later on, in 2016, the Paris Agreement on climate change entered into force. This Agreement specified the obvious requirement to reduce the rise of global temperatures (UN).

For sustainable development to be achieved, it is crucial to harmonize three interconnected core elements: economic growth, social inclusion and environmental protection. The optimistic news are that every year, more affordable and reliable solutions appear and thus people are able to turn into cleaner, more resilient economies. People realize the seriousness of the critical state of the world environment and undertake measures to speed up the changes towards renewable energy and a range of other measures to reduce emissions and increase adaptation efforts (UN).

People have a wide range of a variety of choices of potential adaptation and mitigation pathways. It is however obvious, that all these choices are associated with synergies and trade-offs, which have impact on people's lives (Denton et al. 2014; Fazey et al. 2016).

We expect that the adaptation impacts will be mostly positive for sustainable development, and thus on the SDGs as well. Agriculture and health issues are among especially privileged sectors due to ecosystem-based adaption. There may occur negative impacts of such transformations, with hidden trade-offs in adaptation pathways and risk of reinforcing of the existing inequalities, potentially leading to lock-ins and poverty traps (Fazey et al. 2016).

Pursuing climate mitigation measures generates benefits on a number of various sustainable development levels, and advance short-term targets under the SDGs. Best available technologies to increase efficiency in the use of resources, at all levels, from local to global, help to advance toward the 1.5 °C target, while the technological and behavioral changes will help to recognize the full range of mitigation (Denton et al. 2014).

In this book we provide scientific basis for a number of modern approaches and state-of-the art methods of monitoring both the environment, social behavior and human expectations towards the protection of environment. The authors of the papers, young scientists, discuss these issues from different research perspectives, from physics to social sciences showing the challenges and future scenarios based on the scientific evidence.

References

Denton F et al (2014) Climate-resilient pathways: adaptation, mitigation, and sustainable development. Climate change 2014: impacts, adaptation, and vulnerability. Part A: global and sectoral aspects. In: Field CB et al (eds) Contribution of working group II to the fifth assessment report of the intergovernmental panel of climate change. Cambridge University Press, Cambridge, United Kingdom and New York, NY, USA, pp 1101–1131

Fazey I, Wise R, Lyon C, Câmpeanu C, Moug P, Davies T (2016) Past and future adaptation pathways. Clim Dev 8:26–44. https://doi.org/10.1080/17565529.2014.989192

http://www.un.org/sustainabledevelopment/

IPCC—Intergovernmental Panel on Climate Change (2013) Climate change 2013, the physical science basis. In: Contribution of working group I to the fifth assessment report of the intergovernmental panel on climate change. Cambridge Univ. Press, Cambridge, UK/New York, USA, 1535 p. http://dx.doi.org/10.1017/CBO9781107415324

Pelling M, O'Brien K, Matyas D (2015) Adaptation and transformation. Clim Change 133(113–127):4. https://doi.org/10.1007/s10584-014-1303-0

Ruwa R, Simcock A, Bebianno M, Calumpong H, Chiba S, Evans K, Kamara O, Marschoff E, McClure M, Mohammed E, Park Ch, Randrianarisoa L, Sanchez M, Strati A, Tuhumwire J, Ca Vu T, Wang J, Zielinski T (2017a) The ocean and the sustainable development goals under the 2030 agenda for sustainable development; a technical abstract of the first global integrated marine assessment. United Nations, 35 p

Ruwa R, Simcock A, Bebianno M, Calumpong H, Chiba S, Evans K, Kamara O, Marschoff E, McClure M, Mohammed E, Park Ch, Randrianarisoa L, Sanchez M, Strati A, Tuhumwire J, Ca Vu T, Wang J, Zielinski T (2017b) The impacts of climate change and related changes in the atmosphere on the oceans; a technical abstract of the first global integrated marine assessment. United Nations, 15 p

Urban Green Space as a Tool for Cohesive and Healthy Urban Community

Joanna Stępień

Abstract Urban green spaces play a significant and varied role for preserving high quality of urban life in increasingly urbanized societies. They provide wide range of advantages that support health and subjective well-being of city residents. In many cases, however, these benefits are not fairly distributed across diverse urban populations. This paper presents an overview of the role of urban greenery for city dwellers and discusses the shortage of similar studies in post-transition countries of Central and Eastern Europe, on the example of Poland. The main aim is to take a look on how urban natural settings can be planned and managed in ways that are more meaningful to different groups of people. The results of different studies provide us with knowledge on how to plan and design green spaces for various demographic and socio-economic cohorts in modern cities, although there is still a significant deficiency of dedicated researches, particularly in less developed countries.

Keywords Urban green spaces · Subjective well-being · City residents' health

Green spaces have been founded in cities since the beginning of their existence, initially to ensure nourishment security and for aesthetic reasons. At a later stage, sanitary benefits closely related to urban green settings became increasingly important. However, rapid economic development and corresponding population growth observed in late XX century accelerated public investments, particularly in high-developed regions and dense cities. As a result, decision-makers prioritised economic growth over green spaces as they responded to new socio-economic pressures. Protecting or promoting urban green areas was unsuccessful because their contribution to economic production was greatest when they were sold and developed (Kumagai et al. 2014). Nowadays, facing the dominant economic thinking we are under a pressure to justify the existence of green spaces in cities by

J. Stępień (✉)
Department of Economic Geography, Faculty of Oceanography and Geography, University of Gdansk, Gdansk, Poland
e-mail: joanna.stepien@ug.edu.pl

T. Zielinski et al. (eds.), *Interdisciplinary Approaches for Sustainable Development Goals*, GeoPlanet: Earth and Planetary Sciences,
https://doi.org/10.1007/978-3-319-71788-3_2

using economic arguments. Many recent discussions on human-nature relationships have adopted the perspective of ecosystem services which emphasize that nature provides people with many benefits. Recent reviews of the literature have shown that urban green spaces are indeed important for city dwellers' health and well-being and environmental sustainability, although the specific mechanisms for these benefits are often complex (Kabisch et al. 2015).

1 Urban Green Space, Well-Being and Health

Currently, there is no universally accepted definition of urban green space. This concept may include all places with 'natural surfaces' or 'natural settings', but it also comprises specific types of urban greenery, such as street trees and 'blue areas' which represent water elements ranging from ponds to coastal zones (World Health Organisation, 2016). In broad terms we can treat urban natural areas (green and blue) as all natural areas of public value, both terrestrial and aquatic, which offer important opportunities for sport and recreation and also function as visual amenity (urban forests, parks, squares, beaches, sea, lakes, rivers etc.). The concept of urban green and blue space system emphasizes the quality and the quantity of urban and peri-urban green and blue spaces (Turner 1996; Rudlin and Falk 1999), their multifunctional role (Sandström 2002), and their role in connecting different environments (Van der Ryn and Cowan 1996). The introduction of urban green and blue system as a coherent planning entity creates a framework for economic growth, nature conservation and public health promotion in urban development (Walmsley 2006; Schrijnen 2000; Van der Ryn and Cowan 1996). In policy terms, it is important to focus on urban green space that is open to the public particularly when considering universal green space access for all urban residents, regardless of socioeconomic circumstances. Identifying and assessing the potential of natural space and determining its suitability for meeting the needs of city residents who are users of those spaces correspond with the idea of valorization (Węsławski et al. 2009).

Environment and health are main areas (apart from financial situation, job, leisure, and housing) that constitutes the concept of subjective well-being referring to how people experience the quality of their lives based on their emotional reactions and cognitive judgments (Van Praag et al. 2003; Dennis and James 2017). Modern scholars on well-being and happiness believe those concepts to be similar, and they can also be used in empirical studies on the level of community health (Czapiński 1992, 2000; Persaud 1998). Health as a welfare is sometimes identified with a sense of good quality of life, though this idea is much broader.

Subjective well-being is positively associated with health and people who rate their general health as good tend to experience better self-reported well-being to those who rate health as poor (Okun et al. 1984). However, the idea of health itself is ambiguous and difficult to define. In the colloquial meaning health means the absence of disease or ailment. In professional terms, i.e. formulated by representatives of various scientific disciplines, there is no homogenous conceptualization of health (Piątkowski 2002).

The World Health Organization indicates a broad definition of health as full physical, mental and social well-being. Currently this definition from 1948, is criticized as too extensive which leads to ambiguous interpretation (Heszen and Sęk 2008). More specific concept treats health as a value. In this concept, health and disease, similarly to other values are considered separable. Health has a positive value which people seek for, while disease has a negative value, which people try to avoid. Understanding health as values is particularly useful in health promotion, because it helps to explain why people take actions to protect their health (Gniazdowski 1990).

Taking action aimed at health improvement is likely when it is treated as autotelic value, that means being important itself regardless of other values. However, for some people, health is an instrumental value, which is only a means to achieve other goals. In this context another values may be more important and will be reached at health expense (Heszen and Sęk 2008). Understanding the mechanism of treating health as a value can be useful in explaining recreational and sport activities (pro-health behaviors) taken by city residents, in the context of urban natural spaces usage and specifics of social structure. Social scientists refer also to the concept of health as a lifestyle that reflects specific position occupied by the individual in the social structure and its economic status. Lifestyle, and therefore preferences and choices made by people and their behaviours relate to ways of spending free time and taking care of their own health (Ostrowska 1997, 1999, 2000).

Within the social sciences (see Gniazdowski 1995; Ostrowska 1997, 1999) the issue of lifestyle is considered to be very complex. According to sociological approach, lifestyle is a reflection of belonging to a social class and an expression of socio-political, economic and cultural system. People represent through their lifestyle social and cultural influences. Human preferences, apart from social stratum and economic status, control patterns of spending leisure time. Thus, health-related behavior is a reflection of socialization process and consists of habits positively or negatively associated with health.

2 The Role of Urban Natural Settings for City Dwellers' Health and Subjective Well-Being

Urban natural spaces are of strategic importance for maintaining a high quality of life in increasingly urbanized societies (Chiesura 2004). The essential contributions of nature to the quality of urban life include the generally known environmental benefits of to the air, water, soil, and the ecosystem as a whole (Chiesura 2004), the psychological and physical benefits of reduced stress to citizens (Kuo et al. 1998) through the encouragement of physical activity and the extension of elderly people's life span (Takano et al. 2002), and the social and economic benefits to the city of increasing social integration and interaction among neighbours (Coley et al. 1997). Urban natural spaces play also an important role from a social perspective by promoting a sense of safety, social support and cohesion, and integration (Coley et al. 1997; Jay and Schraml 2009; Kuo 2003; Maas et al. 2009; Peters et al. 2010). Due to their unique benefits, natural areas have been recognized as one of the most important components of urban areas.

Although the evidence remains mixed (see Badland et al. 2010), some studies show that the availability of natural spaces disproportionately benefits more affluent communities. Other researches have shown more socioeconomically deprived areas and areas with high ethnic minority populations may have poorer quality natural areas (Crawford et al. 2008; Pearce et al. 2007; Timperio et al. 2007), which in turn amplify social inequalities. The link between an individual's socio-economic position and their health is well-established (Bartley et al. 1997; Brunner 1997; 1997). Furthermore, epidemiological studies have provided evidence of a positive relationship between longevity and access to green space (Takano et al. 2002; Tanaka et al. 1996), and between green space and self-reported health (De Vries et al. 2003). Those analyses form a joint, broad research trend referring to the issue of urban environmental justice.

Despite the growing research in this area, there is comparatively little evidence demonstrating differential health benefits associated with specific characteristics of green space. Varying configurations of green space, built environment and topographical features near a person's residence may offer different opportunities for physical activities and mental renewal, depending on the person's age, gender and individual preferences (Wheeler et al. 2015). An urban green space may have varying qualities that offer different opportunities for relaxation, engagement with the natural environment, physical exercise and athletic activities or getting away from unpleasant aspects of the urban environment, such as noise or heat.

Douglas et al. (2017) review the evidence linking health, well-being and green space using a life course approach that gives the guidance for the provision of more inclusive green spaces which respond to the varying needs of people across all life-course stages. According to their research the childhood interactions with and within green spaces are beneficial for the health of children both physical and psychological, as well as for their social ad intellectual development. The design of parks which promote physical and social well-being in teenagers emerges as a potentially key focus in promoting life-long physical and psychological health and well-being through childhood, adulthood and old age. Planning and urban design can facilitate green space activity and recreation among older people by providing accessible and safe green spaces with well-maintained walking infrastructure, that is safe and wheelchair accessible. Such provision can encourage older people to observe, use and benefit from public green space for as long as their health condition allows.

Urban natural spaces provide a wide range of benefits, or ecosystem services, that support physical, psychological, and social health. In many cases, however, these benefits are not equitably distributed across diverse urban populations.

3 The Shortage of Researches on Urban Greenery and Health in Poland

Most of the studies on the nature-human relations and urban ecosystem services have been conducted in Western European countries and North America (Haase et al. 2014). Some findings, such as the stress reduction opportunities that urban green space affords, have been replicated in multiple studies conducted in different countries. However, most of the epidemiological studies have been carried out in high income countries (mainly in western and northern Europe, as well as in North America, Australia and Japan).

There is broad literature on the significance of nature for inhabitants' health and quality of life but most of the researches were realised in high developed countries and regarded epidemiological health measures based on statistical health data. Although some researches have shown that public parks and green space provide a variety of physical, psychological, and social benefits to urban residents, few studies have examined the influence of parks on comprehensive measures of subjective well-being at the city level (Larson et al. 2016). Moreover, very little is known about the (additional) positive effects of green space on wellbeing through mechanisms of increased and prolonged physical activity, and improved social cohesion (Maas et al. 2006). Some evidence suggests that exposure to local green space confers a health benefit on more general level (for example country or group of countries) but this association is not observed at the city level. Further work is needed to establish how urban residents interact with 'local' green space, in order to ascertain the most relevant measures of green space for improving their quality of life and well-being (Bixby et al. 2015).

World Health Organisation emphasizes the need for more research on urban green space and health in the eastern parts of Europe. Such research is essential for assessing health benefits of urban green spaces in middle and low income countries and in cities with different urban design characteristics (World Health Organisation, 2016). Poland, similarly to other Central and East European countries, has not been the area of such studies on the role of urban natural areas for city residents' health and well-being. Particularly the literature on the governance of urban ecosystem services has been clearly biased towards some regions, and while Western Europe has been well represented, the post-transition countries remain almost a blank spot (Kronenberg 2015).

Choices and preferences of people, built in stable, old western democratic countries cannot be directly transferred to a post-communist country—as for example was demonstrated in consumer choices with price, and not quality, being the main factor in shopping (Degeratu et al. 1999). Similar contradiction relates to health which as a declared value has high position in the hierarchy of values, but in practice healthy lifestyle and preventive health care is rarely implemented in Polish society (Bogusz 2004).

Polish economic and social situation differs from Western European countries. Changes in Poland as a result of political transformation in the 1980s are more rapid than average, but there is still no clear established middle class in the social structure. Moreover, the characteristic feature of Polish social hierarchy is little mobility measured by changes in the position and status in subsequent generations. This means that inequalities reproduce themselves, and certain groups remain privileged, as for example with access to education. Despite the "revolutionary" political and economic changes and political elite transition after 1989, at a deeper level of social structure still the same mechanisms exist as in the final period of the Polish People's Republic (Domański 2009). This factor may be crucial while examining peoples' value perception, choices and preferences taking Polish social structure into consideration.

Another significant difference between West and East is the level of urbanization in Poland which is clearly lower (60%) than in Western Europe (80%). However, as a result of rapid political and economic changes which happened after 1989, many big cities transformed their composition due to massive housing or other infrastructure constructions, interfering with natural environment within cities. As a result, natural spaces were very often modified and lost their primary functions.

Although during the economic and social transition period in Poland several environmental indicators in cities have improved (e.g. emissions of pollutants from industry; share of treated waste water), other problems have emerged or intensified (e.g. increased generation of waste and traffic) (Kronenberg and Bergier 2012). The increased wealth and new socio-cultural patterns have led to further urbanization, growth of urban populations, and urban sprawl. Meanwhile, the urban structure in city centers in many cities has been worsening (degradation of old buildings and urban greenery, increasing areas covered by roads and car parks). As a result, pressure on urban ecosystems has been increasing both within city centers and in the greater city areas, resulting in shrinking green areas and increasing areas covered with paved surfaces (Kabisch, Haase 2013).

Paradoxically, greenery was frequent in socialist cities, and it only started to shrink rapidly with the economic transition. The neoliberal economic system entailed the extensive privatization of land and the allocation of public space for private uses, lifting the relatively strict socialist-era building regulations, and limiting investment in public goods, instead favouring what the market demanded (i.e. in the interest of the wealthier spheres of society) (Hirt 2013). The role of the public sector was minimized, but it was still expected to manage public space with the limited funds available for these purposes.

The use of resources for present purposes may endanger future resilience of cities what we can observe even in high developed countries like Japan. Green spaces in Tokyo provided in the past a flexible, but gradually disappearing resource. The adaptive capacity secured by the green space resources is currently no longer available in the capital city. The Tokyo case underscores the risk inherent in the reduction of non-renewable resources to accommodate economic growth and short-term resilience at the expense of long-term resilience (Kumagai et al. 2014).

Lack of awareness of the significance of urban green infrastructure more generally, is definitely an important problem that affects proper management of urban ecosystem services. It is reflected in many institutional failures that collectively explain resistance, or at least apathy, to urban greening. However, these institutional failures are also related to other important problems, such as institutions being locked in historical trends favouring other interests over the protection of green areas. The poor quality of institutions has been one of the most important barriers to sustainability in Poland and in other transition and post-socialist countries (Kronenberg and Bergier 2012), where the sudden collapse of the previous political system and the subsequent focus on economic transition, the free market and consumption left little space for other issues.

Meanwhile, the public sector in post-socialist countries was (and still is) affected by many institutional problems, especially when compared with the situation of its counterpart in Western countries. The common problems include lack of supplies of green areas (which makes planning and management difficult), low level of innovativeness in terms of acquiring funds for urban forest management, ad hoc management activities (not part of a comprehensive and strategic approach to the management of green areas), poor communication between different departments, and poor public participation processes (Gudurić et al. 2011). These issues have also been confirmed in Poland.

Structural problems result in limited opportunities for investment in environmental protection. Disproportionately more attention has been paid to basic economic concerns than to environmental objectives, including many solutions that favoured investment in public goods. These problems translate into a lack of creativity for problem solving in the public sphere and over trust on the existing institutional structures, as opposed to openness to new ideas or institutional reform. They illustrate the difficulties in getting local authorities responsible for urban greenery to adopt new management solutions, as they are confronted by the 'other priorities' and the conviction that there is no space for greenery in modern city centers.

The main need to research in Poland is to identify and define the role of urban natural areas (land and water) for health and subjective well-being of urban residents, in the context of changing social structure after economic transformation. The research should verify whether the availability of natural areas in the city is a limited luxury only to a privileged part of the society or of vital importance for the city dwellers, that is necessarily to maintain a high quality of life and self-assessment of health by residents, regardless the city social structure. The modifying role of the social structure on the perception of health and well-being of citizens and assigning the value to natural areas in the city has not been widely studied in Poland, and the results of similar studies conducted in other countries are ambiguous.

Considering the complexity of main research problem we can formulate a number of specific research questions that examine the general people attitude toward natural resources and areas within city, their opinions on values given to nature, perception on the accessibility, quantity and quality of spaces for recreation and spending leisure time, and general pattern of health behaviors (sport, recreation):

- What are the social and environmental determinants of subjective well-being of city dwellers?
- How and to what extent differences in the social structure of urban population affect the perception and usage of natural spaces?
- To what extent the quality, quantity and availability of natural resources determine health behaviors (physical activity, recreation and leisure activities) of residents?
- How are green spaces distributed and utilized across different communities?

Those general issues allow to formulate more specific and detailed questions on the role of nature for subjective well-being and find out the mechanism of taking physical activities within natural areas in urban environment depending on place within social structure:

- How are cultural ecosystem services perceived and valued among different populations at both the community and household scale?
- What characteristics of urban green space maximize the delivery of cultural services which support social determinants of health?
- Does a direct access to more natural urban areas have an influence on residents' decisions on the place of dwelling?
- Through which pathways can natural spaces precipitate specific health outcomes?
- Are there strong traditional bonds to some forms of nature use?
- How those aspects correspond with the self-assessment of health?
- To what extent the amount and quality of green and blue areas can be employed as a key explanatory factor in relation to subjective well-being?

Recognizing urban residents' perception on the value of natural areas and their behaviors has both cognitive and practical aspect. In the cognitive sense it will be interesting to examine the attitudes and values attributed by city residents to health, their health behaviors (recreation, sport, leisure time etc.), and the usage of natural areas. In practical sense it is important to find out to what extent natural urban areas influence residents' subjective well-being and their quality of life.

4 Conclusion

This paper has presented an overview of the role of urban natural spaces for city residents and discussed the shortage of similar studies in post-transitional countries in Central and Eastern Europe, on the example of Poland. The aim was to take a look at the ways on how urban green spaces can be planned and managed in ways that are more meaningful to different people and groups.

Urban planners, managers and policy-makers face conflicting demands to promote more compact cities with greater population density in order to create critical mass to support and justify public and private service provision. This also applies to provision of green space and improvement of urban ecosystem services. The current studies show that urban green space has health benefits, particularly for economically deprived groups, children, pregnant women and senior citizens. It is therefore essential that all communities have adequate access to green space, with particular priority placed on provision for disadvantaged groups. While details of urban green space design and management have to be sensitive to local conditions, the need for green space and its value for health and well-being is universal. A city of well-connected, attractive green spaces that offer safe opportunities for urban residents for active mobility and sports as well as for stress recovery, recreation and social contact, is likely to be more resilient to extreme environmental events, such as heat or extreme rainfall. Such a city is also likely to have healthier citizens, reducing demands on health services and contributing to a stronger economy.

Urban growth, by changing cities and the surrounding countryside, presents numerous challenges for the maintenance of urban green space, and consequently also for human health and well-being. In this regard, the studies on Polish specifics can contribute to better understand the mechanisms that lead to positive outcomes for people of different socio-economic statuses, and thereby assist urban planners and decision makers in designing more successful strategies for socially, economically and environmentally sustainable settings. At the same time, it would be important to include residents in the planning process and consider their needs and expectations, as they are the final users of these settings. Conducting researches on the relation between urban natural setting and city dwellers in Poland can allow to use limited and declining urban natural areas more efficiently and adapt them to various needs of their residents. In the long-term perspective it may improve health and subjective well-being of city inhabitants, and as a result increase their quality of life.

References

Badland H, Keam R, Witten K, Kearns R, Mavoa S (2010) Examining public open spaces by neighbourhood level walkability and deprivation. J Phys Act Health 7(6):818–824

Bartley M, Blane D, Montgomery S (1997) Socio-economic determinants of health: health and the life course: why safety nets matter. BMJ 314:1194–1203

Bixby H, Hodgson S, Fortunato L, Hansell A, Fecht D (2015) Associations between green space and health in English cities: an ecological, cross-sectional study. PLoS ONE 10(3):1–12

Bogusz R (2004) Zdrowie jako wartość deklarowana i realizowana. [w:] W. Piątkowski (red.). Zdrowie – choroba – społeczeństwo. Studia z socjologii medycyny. Wyd. UMCS, Lublin, pp 127–139

Brunner E (1997) Socio-economic determinants of health: stress and the biology of inequality. BMJ 314:1472–1485

Chiesura A (2004) The role of urban parks for the sustainable city. Landscape Urban Plan 68 (1):129–138

Coley RL, Kuo FE, Sullivan WC (1997) Where does community grow? The social context created by nature in urban public housing. Environ Behav 29(4):468–494

Crawford D, Timperio A, Giles-Corti B, Ball K, Hume C, Roberts R (2008) Do features of public open spaces vary according to neighbourhood socioeconomic status? Health Place 14 (4):889–993

Czapiński J (1992) Psychologia szczęścia. Przegląd badań i zarys teorii cebulowej. Oficyna Wydawnicza Akademos, Warszawa

Czapiński J (2000) Stan zdrowia psychicznego – wskaźniki niekliniczne. [w:] J. Cz. Czabała (red.). Zdrowie psychiczne. Zagrożenia i promocja. Instytut Psychiatrii i Psychologii, Warszawa

Davey-Smith G, Hart C, Blane D, Gillis C, Hawthorne V (1997) Lifetime socio-economic position and mortality: further evidence from the Whitehall study. J Epidemiol Commun Health 44:260–265

De Vries S, Verheij RA, Groenewegen PP, Spreeuwenberg P (2003) Natural environments—healthy environments? Environ Plann 35:1717–1731

Degeratu AM, Rangaswamy A, Wu J (1999) Consumer choice behavior in online and traditional supermarkets: the effects of brand and name, price and other search attributes. Int J Res Mark 17(1):55–78

Dennis M, James P (2017) Evaluating the relative influence on population health of domestic gardens and green space along a rural-urban gradient. Landscape Urban Plan 157:343–351

Domański H (2009) Stratyfikacja a system społeczny w Polsce. Ruch prawniczy, Ekonomiczny i Socjologiczny 2:381–395

Douglas O, Lennon M, Scott M (2017) Green space benefits for health and well-being: a life-course approach for urban planning, design and management. Cities 66:53–62

Gniazdowski A (red) (1990) Zachowania zdrowotne. Zagadnienia teoretyczne, próba charakterystyki zachowań zdrowotnych społeczeństwa polskiego. Instytut Medycyny Pracy im. Prof. dr med. Jerzego Nofera, Łódź

Gniazdowski A (1995) Rola wybranych czynników psychospołecznych w procesie promocji zdrowia w Polsce. Promocja Zdrowia Nauki Społeczne i Medycyna 2:22–28

Gudurić I, Tomićević J, Konijnendijk CC (2011) A comparative perspective of urban forestry in Belgrade, Serbia and Freiburg, Germany. Urban For Urban Green 10:335–342

Haase D, Larondelle N, Andersson E, Artmann M, Borgström S, Breuste J (2014) A quantitative review of urban ecosystem service assessments: concepts, models, and implementation. Ambio 43:413–433

Heszen I, Sęk H (2008) Psychologia zdrowia. Wyd. PWN, Warszawa

Hirt S (2013) Whatever happened to the (post)socialist city? Cities 32:29–38

Jay M, Schraml U (2009) Understanding the role of urban forests for migrants—uses, perception and integrative potential. Urban For Urban Greening 8(4):283–294

Kabisch N, Qureshi S, Haase D (2015) Human–environment interactions in urban green spaces—a systematic review of contemporary issues and prospects for future research. Environ Impact Assess Rev 50:25–34

Kronenberg J (2015) Why not to green a city? Institutional barriers to preserving urban ecosystem services. Ecosystem Services 12:218–227

Kronenberg J, Bergier T (2012) Sustainable development in a transition economy: business case studies from Poland. J Clean Prod 26:18–27

Kumagai Y, Gibson R, Filion P (2014) Evaluating long-term urban resilience through an examination of the history of green spaces in Tokyo. Local Environ Int J Justice Sustain, 1–22

Kuo FE (2003) The role of arboriculture in a healthy social ecology. J Arboric 29:148–155

Kuo FE, Sullivan WC, Coley RL (1998) Fertile ground for community: inner-city neighbourhood common spaces. Am J Community Psychol 26(6):823–851

Larson RL, Jennings V, Cloutier S (2016) Public parks and wellbeing in urban areas of the United States. PLoS ONE 11(4):1–19

Maas J, Verheij RA, Groenewegen P, de Vries S, Spreeuwenberg P (2006) Green space, urbanity, and health: how strong is the relation? Epidemial Community Health 60:587–592

Maas J, Van Dillen SME, Verheij RA, Groenewegen PP (2009) Social contact as a possible mechanism behind the relation between green space and health. Health Place 15(2):586–595

Okun MA, Stock WA, Haring MJ, Witter RA (1984) Health and subjective well-being: a meta-analysis. Int J Aging Hum Dev 19(2):111–132

Ostrowska A (1997) Prozdrowotne style życia. Promocja zdrowia Nauki Społeczne i Medycyna 4:7–24

Ostrowska A (1999) Styl życia a zdrowie. Z zagadnień promocji zdrowia. Wyd. Instytutu Filozofii i Socjologii PAN, Warszawa

Ostrowska A (2000) Społeczne czynniki warunkujące zachowania prozdrowotne – bilans dekady. Promocja Zdrowia. Nauki Społeczne i Medycyna 19:46–65

Pearce J, Witten K, Hiscock R, Blakely T (2007) Are disadvantaged neighbourhoods deprived of health-related community resources? Int J Epidemiol 36:348–355

Persaud R (1998) Pozostać przy zdrowych zmysłach. Jacek Santorski & Wydawnictwo, Warszawa

Peters K, Elands B, Buijs A (2010) Social interactions in urban parks: stimulating social cohesion? Urban For Urban Greening 9(2):93–100

Piątkowski W (2002) Zdrowie jako zjawisko społeczne, [w]. In: Barański J, Piątkowski W (red.) Zdrowie i choroba. Wybrane problemy socjologii medycyny, Wrocławskie Wydawnictwo Oświatowe, Wrocław, p 35

Rudlin B, Falk N (1999) Building the 21st century home. The sustainable urban neighbourhood. Architectural Press, Oxford

Sandström UF (2002) Green infrastructure planning in urban Sweden. Plann Pract Res 17 (4):373–385

Schrijnen PM (2000) Infrastructure networks and red-green patterns in city regions. Landscape Urban Plann 48:191–204

Takano T, Nakamura K, Watanabe M (2002) Urban residential environments and senior citizens' longevity in megacity areas: the importance of walkable green spaces. J Epidemiol Community Health 56(12):913–918

Tanaka A, Takano T, Nakamura K, Takeuchi S (1996) Health levels influenced by urban residential conditions in a megacity—Tokyo. Urban Stud 33:879–894

Timperio A, Ball K, Salmon J, Roberts R, Crawford D (2007) Is availability of public open space equitable across areas? Health Place 13:335–340

Turner T (1996) City as landscape. A post-postmodern view of design and planning. E&FN Spon, London

Van der Ryn S, Cowan S (1996) Ecological design. Island Press, Washington, DC

Van Praag BMS, Frijters P, Ferrer-i-Carbonell A (2003) The anatomy of subjective well-being. J Econ Behav Organ 51:29–49

Walmsley A (2006) Greenways: multiplying and diversifying in the 21st century. Landscape Urban Plann 76:252–290

Węsławski JM, Warzocha J, Wiktor J, Urbański J, Radtke K, Kryla L, Tatarek A, Kotwicki L, Piwowarczyk J (2009) Biological valorisation of the southern Baltic Sea (Polish Exclusive Economic Zone). Oceanologia 51(3):415–435

Wheeler BW, Lovell R, Higgins SL, White MP, Alcock I, Osborne NJ, Depledge MH (2015) Beyond greenspace: an ecological study of population general health and indicators of natural environment type and quality. Int J Health Geographics 14:1

Examples of Innovative Approaches to Educate About Environmental Issues Within and Outside of Classroom

Izabela Kotynska-Zielinska and Martha Papathanassiou

Abstract In this paper we describe innovative approaches to educating about environmental issues, with the main focus on marine environment. We discuss the importance of non-formal education in the learning process, with the main focus on oceans and the marine environment. We present examples of successful projects which resulted in the creation of concrete platforms for further education, such as the Blue Society [Sea for Society EU Project (SFS)] or I Live by the Sea, international photo contest by Today We Have. The examples of the non-formal education activities which have been described in the paper provided all participants, including our teams, with a mutually valuable experience which provided important contribution for future collaboration with schools and life marine educational programs. We succeeded in increasing environmental awareness, as well as commencing an exchange of information and a discussion of controversial issues regarding marine environments among the participants from different parts of the world.

Keywords Non-formal education · Marine issues · SDGs · I Live by the Sea 10 × 20 Initiative

1 Introduction

It is widely accepted that interest in science, especially among young people, is declining. This decline results in the confusion between science, pseudoscience and "junk science" (Kotynska-Zielinska et al. 2016). Given increased public participation in decision-making, the limited understanding of science could compromise evidence-based decision making in the long run (Bray et al. 2012).

I. Kotynska-Zielinska (✉)
Today We Have, Sopot, Poland
e-mail: kotynska-zielinska@todaywehave.com

M. Papathanassiou
Indigo-Med, Athens, Greece

T. Zielinski et al. (eds.), *Interdisciplinary Approaches for Sustainable Development Goals*, GeoPlanet: Earth and Planetary Sciences,
https://doi.org/10.1007/978-3-319-71788-3_3

It is a common practice that science is presented out of context and thus it seems irrelevant to peoples' lives. As a result, many of them lose interest. If a person's own motivation is disrespected, even the most careful preparation on the part of the educator will lead to failure (Kuhn 1993). It is, therefore, crucial to highlight the importance of science and its relevance to peoples' lives (George and Lubben 2002; Lubchenco and Grorurd-Colvert 2015; Stocklmayer and Bryant 2012). People, especially young learners, also need more positive and realistic demonstrations of the scope and limitations of science and scientists (Ainley and Ainley 2011; Anderhag et al. 2013).

Oceans are a vital element of planet Earth and humankind (The Convention on Biological Diversity's High-Level Panel…). Anything which is important to us is related to the oceans, from the air we breathe, to various uses of drinking water, food, entertainment. It comes from or can be shipped by the ocean. No matter how far from the shore people live, oceans affect their everyday lives (Costanza et al. 1997; Costanza 1999; Beaumont et al. 2007; Wallmo and Edwards 2008).

However, oceans are in a severe and deepening crisis, threatening the environment and human well-being (Weslawski et al. 2006; Ronnback et al. 2007). The first among the problems that are plaguing the ocean is overfishing, which has caused the collapse of fish stocks and the loss of biodiversity. This is a global problem and it will require a global solution to bring the ocean back from the brink of irreversible decline (Brander et al. 2015; Lubchenco and Grorurd-Colvert 2015), currently, the most effective ways to create such a solution are beginning to emerge from international collective actions starting at the United Nations (UN) (http://www.un.org/sustainabledevelopment/).

One of the actions which has been undertaken is the 10 × 20 Initiative, a plan of actions to develop a scientifically-based framework to assist Member States in achieving the globally agreed commitment to conserve at least 10% of coastal and marine areas by the year 2020 (http://www.oceansanctuaryalliance.org/). This is Target 5 of the UN's SDG 14: Conserve and sustainably use the oceans, seas and marine resources (http://www.un.org/sustainabledevelopment/). The aim of the 10 × 20 Initiative is to create a thought out, coordinated global network of Marine Protected Areas (MPAs) that achieves the 10% target to conserve biodiversity (http://www.oceansanctuaryalliance.org/).

Today, researchers from universities and research institutions are expected to provide more than just scientific results, which are very often not "understandable" for the general public. This new situation has led to an open dialogue between science and society (http://seaforsociety.eu). However, it is a difficult path and both the scientists and the public must learn how to communicate, especially since raising awareness and developing a culture that values scientific endeavor must come from non-formal, informal and formal education, namely by scientists, the general public, schools, parents and young learners.

It is possible to increase the understanding of science through mobilization activities, and through active participation of the general public "in" science (http://seaforsociety.eu). However, before adequate empowerment strategies are developed, it is necessary to reconstruct the current perception of a given scientific phenomenon.

We realize the fundamentals of modern education, i.e. key elements such as: any kind of learning is valuable, learning is a continuous challenge and that any kind of interdisciplinary approaches create more opportunities in learning than the use of one type of approach. Therefore, in this work, we present selected cases of non-formal types of activities in relation to the marine environment. We describe a number of local and international initiatives that promoted the awareness of marine issues among young learners. We focus mainly on the fact that these activities were implemented both within and outside of the school system.

2 The Importance of Science Education in the Modern World

According to the European Commission, knowledge of and about science is crucial in producing active, creative and responsible citizens in the modern society, given that citizens equipped with such skills are not only aware and innovative but also able to work collaboratively to face the complex challenges of the modern world (European Commission 2015). Such citizens are able to better understand our world, and thus to follow and even guide its development as well as to wisely plan its future. Scientific literacy also places citizens in an important part of the global culture. There are a number of key issues which make science education vital to society (European Commission 2015). Some of these include:

- Promotion of culture for critical, science based, thinking to inspire citizens to use facts as the basis for decision making;
- Providing citizens with an appropriate level of knowledge and skills and thus confidence, to be active players in an increasingly complex reality;
- Inspiring young learners at all levels and of various talents to be creative and aspire to study science and further seek careers in science and other areas and professions that require current knowledge and create opportunities to plan the future.

Therefore, it is obvious that science education should be in the center of broader educational goals for the general society. However, science education, which is not properly balanced with expectations and requirements may lead to the demotivation of learners of all ages. In such a case, learners may feel "pushed away" from further endeavors, at all educational levels, up to later stages in their lives which often results in a difficulty to make proper, scientifically based decisions. Contemporary science education should focus on competences, and not on learning science, but learning *through* science. Another strong aspect of a modern approach to teaching science is to increase synergies between science, creativity, and innovation. The modern education paradigm facilitates cooperation among schools and other stakeholders, with a very special role played by the learners' families. In such a model of education, including science education, it is natural that representatives of the general society, professionals of a variety of fields of expertise, become actively involved in the educational system by providing "real-life" projects, and the engagement of learners.

3 Examples of Good Practices of Marine Education Initiatives

3.1 Sea for Society (http://seaforsociety.eu)

The results of previous surveys (e.g. FP5 OCEANICS project or The Ocean Project) have shown that there is a serious gap in the knowledge of European citizens related to the daily link they have with marine ecosystem services. Citizens are not fully aware of how they benefit from these services nor how their everyday lives potentially affect them.

Sea for Society (SFS) was a European Project funded by DG Research & Innovation under the theme: Science in Society. The idea of the project was to engage various stakeholders, citizens and youth in an open and participatory dialogue to share knowledge, forge partnerships and empower actors on societal issues related to the Ocean. One of the most important results of the Project was the development of the concept of "Blue Society".

Within the Sea for Society project, marine researchers and actors, society-oriented organizations and individual citizens, including youth, were brought together in a mutual learning, consultation process and joint action, to work on key issues and develop challenging solutions to strengthen sustainable management of marine ecosystem services by European citizens.

In order to achieve its goals, the SFS team applied the approach based on the Ecosystem Services, which are used within the project to manage a complex marine ecosystem. The Ecosystem Services are related to "Lived Experiences" of stakeholders and citizens, while the Marine Ecosystem Services, describe marine ecosystem based benefits, that people obtain on the daily basis, including a variety of areas from the open seas and oceans, through coastal areas and estuaries.

Avital part of the SFS project was the consultation process, during which the following 6 key issues, relevant to all stakeholders and citizens, were discussed:

- A place to live;
- Energy;
- Food supply;
- Human health;
- Leisure and tourism;
- Transport.

The Sea for Society consultations reveal barriers to a sustainable marine ecosystem and importantly, how stakeholders perceive these barriers to be interrelated and connected. Sea for Society also identified the options considered by stakeholders as most feasible and effective actions for a sustainable marine ecosystem. It also connected the views, experiences and knowledge of stakeholders with the benefits reported by the citizens/youth. The ideas generated during these consultations by the stakeholders and citizens/youth facilitated the development of co-authored actions for mobilization. Sea for Society mobilized stakeholders and citizens/youth for change. Sea for Society actually went beyond the individual campaigns to a collective scaled-up Blue Society movement.

Over a course of 32 SFS stakeholder and citizen/youth consultations across Europe over 500 participants were involved, relatively evenly distributed between both groups. Close to 150 participants from both stakeholder and citizen/youth groups were involved in the Baltic region consultations (Sweden and Poland), while those numbers were around a hundred for each of the two groups that participated in the Atlantic region consultations (Ireland, France and Portugal) and over 180 participants from stakeholder and citizen/youth groups participated in the Mediterranean region consultations (Greece, Spain and Italy).

The stakeholder consultations resulted in close to 800 statements of barriers to a sustainable marine ecosystem and over 650 options to overcome these barriers. These barriers were portrayed on 16 national structural maps, highlighting pathways of aggravation between barriers. The analyses revealed the key sets of barrier themes that were most influential to a sustainable marine ecosystem, for the EU as a whole and across different EU regions. An EU Influence Map and 3 Regional Influence Maps were generated to show the aggravating influence across the 12 higher-order barrier themes.

In the 16 citizen/youth consultations, a total of close to 800 benefits and almost 800 further contributions were generated. These benefits and further contributions/benefits represent needs that are already met and others that have yet to be met by the ocean and its resources. Beside generating benefits and contributions to the 6 described themes, citizens/youth identified contributions to an additional 15 themes (Fig. 1).

The EU Influence Map of Stakeholder Barrier Themes should be read left to right, with the themes on the left having more overall aggravating influence to a sustainable marine ecosystem than barrier themes on the right. This means the Attitudes and Awareness barrier theme, stage one, exerts the highest level of overall

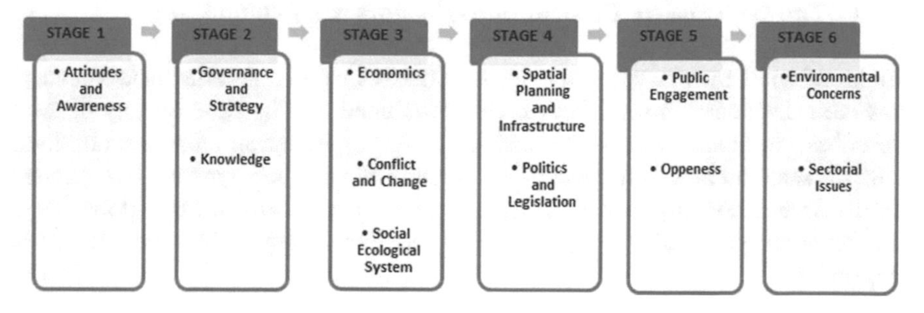

Fig. 1 The Sea for Society multistage EU influence map of stakeholder barrier themes (modified from: Sea for Society)

Table 1 Citizen/Youth top voted benefits and contributions (modified after: Domegan et al. 2016)

Top voted benefits	Top voted contributions
Provision of food (14/15}	Non-renewable energy for transport (10/14)
Transport of goods (12/15)	Diversifying the fish we eat (7/12)
Living by the sea (10/15)	Research and funding (7/12)
Pharmaceuticals/Medical (10/15)	Awareness (7/12)
Well-being and quality of life (10/15)	Relaxation retreats/Spas (7/13)
Passenger transport (9/15)	Eco-tourism/Sustainable tourism (6/14)
Therapy (9/15)	Infrastructure (6/14)
Healthy food (9/15)	Cleaner beaches (6/14)
Fitness/sport (9/15)	
Renewable energy (9/15)	
The sea provides energy (9/15)	
Sustainability (8/15)	
Sport and leisure (8/15)	
Coastal/Urban development (8/15)	

influence to a sustainable marine ecosystem. Attitudes and Awareness according to stakeholders is about "unfounded attitudes and lack of awareness of marine issues". The second highest level of overall influence comes from the Governance & Strategy barrier theme together with the Knowledge barrier theme. Governance & Strategy is seen by stakeholders as "responsible institutional framework strategies, policy and research for marine ecosystems" while Knowledge is about "inadequate scientific and general public knowledge." The stage six barrier themes of Environmental Concerns, "negative marine environment management" and Sectorial Issues, "the disconnect between sector and marine issues" differ from the previous barrier themes and stages, as they have influence, but at the lowest level for a sustainable marine ecosystem (Table 1).

3.1.1 The Blue Society Concept (http://seaforsociety.eu/np4/20/)

As a result of the works within the Sea for Society Project, the Blue Society concept has been developed in relation to ocean sustainability. The Blue Society concept describes the fundamental values and principles of European citizens who "live" with the seas and how, if necessary, this relationship can be improved. It is inspired by the Green Economy concept which in turn has been developed at a global level. The foundations for this concept lay in the fact that society is driven by two basic imperatives:

- To satisfy the needs of our planet's inhabitants today;
- To preserve our natural and manmade resources for future generations.

In the case of the Blue Society, the resources and services are related to those provided by the seas. The Blue Society concept is based on the encouragement of society to take action and it inspires people to actively implement necessary solutions. It focuses on sustainable development policies and a green economy.

The Blue Society concept is tailored to assist stakeholders, researchers and citizens to work together towards reaching "inclusive solutions", since we are all members of a Blue Society.

An important phase of creating the Blue Society was the SFS Mobilization phase which was aimed at a wide range of actors, from the general public and the youth to stakeholders such as business people from the maritime sphere, researchers, educators, and funding agencies.

The Mobilization activities included training for capacity-building in science communication organizations, science museums and/or aquaria, journalists, marine institute communication departments. The approach involved various programs, such as: Blue Talks, Blue Cafés, Living Labs, Youth Parliaments, Open Days, or Festivals. Between 2012 and 2015 over 20 science events had been launched all over Europe.

One example involves the re-launching of the CHARGER boat, developed within the scope of an American project dedicated to the education of young learners about the work of marine scientists and about the seas. Within the SFS Project, this activity was run by two Portuguese partners, Ciência Viva and IST MARETEC.

The CHARGER boat was created by John Winthrop Middle School (Deep River, Connecticut, USA) within the scope of the Educational Passages program. The program was launched in the USA in early 2012 and its goal was to educate young learners about marine environments. The boat travelled across the Atlantic Ocean and landed on 29 January 2014 on the Torreira beach in Portugal.

The boat was found on the beach by a school girl and transported to the IST MARETEC. The boat was renovated and then put back in the Atlantic Ocean on 2 June 2014.

To assist the re-launching, the SFS team ran a number of meetings at the local school to inform the local citizens, including young learners, and get them involved in the re-launch project. As a result, together with the students at the John Winthrop Middle School, the boat is being closely followed by researchers at two Portuguese institutes, the IST MARETEC and the IST Institute for Systems and Robotics. The researchers use the information about the path of the boat to calibrate the MOHID Water Modelling System, an integrated water modelling software developed at the IST, to predict the fate of an instrument drifting at sea using information about sea currents and winds. The boat is equipped with a GPS, and thus it can be followed online at: http://www.nefsc.noaa.gov/drifter/drift_ep_2014_1.html.

This a good example of an activity that facilitates the direct involvement of citizens in science. As a result, citizens become more aware of marine science issues, and are engaged in marine research helping them realize how valuable the marine environment is and what unique resources we are dealing with.

3.2 I Live by the Sea Project

3.2.1 I Live by the Sea, Parts I–III, Local Level (Polish: Mieszkam nad Morzem) (http://todaywehave.com/Wydarzenie_1.html)

This project is an example of non-formal education on marine issues dedicated mostly, but not exclusively, to young learners. In the first three stages of the Project, we started a competition for primary and junior high school students (children aged 7–13) of the Baltic region, near Gdansk, Poland. "I live by the Baltic seaside" was the general theme of the competition, and all three editions involved three different visual arts, i.e., posters and sculptures, movie making, and theatre. In the third edition climate change in the Arctic was added as an additional topic.

In each edition, teams of students were asked to explain—using the selected media—what living by the Baltic Sea (or alternatively what climate change) means for them, and what they find important in this ecosystem. They were also requested to prepare a short presentation on their work, and present it in front of a jury, comprised of natural scientists and artists, as well as teachers and fellow-students. Some of the issues dealt with in all three editions included:

- Overfished oceans and seas;
- How to behave on the beach;
- Biodiversity and invasive species;
- Melting glaciers in the Arctic.

Prior to each edition a number of meetings with scientists were organized for the students. During these meetings, the children were educated on various sea related issues. We learned that setting the theme and giving full freedom of expression to the children was a very successful approach in comparison with previous contests in which participants had to follow strict rules.

Having school teachers involved was a very helpful element of the entire project since they knew their students well and were able to assist them in making the right choices for their presentations. The children conducted the research on the topic out of natural curiosity and were able to provide very detailed information on e.g. invasive species in the Gulf of Gdansk, a potentially serious problem for the regional biodiversity which, however, is not very widely "promoted" among the general public.

Another aspect of the non-formal education of the Project was the creation of a space on the beach for the summer season of July–August 2015 at the beach resort of Sopot, part of the Tri-City agglomeration, one of the most popular summer resorts in the Baltic Sea region. We involved researchers from the Institute of Oceanology Polish Academy of Sciences in Sopot who shared their knowledge with the general public right on the Sopot beach. We had a small area for people to come and discuss issues related to their interests and even concerns, e.g. rapid changes in sea water temperature, cyanobacteria blooms, endangered species, marine protection issues, etc. This allowed us to gain a lot of interesting experiences last year. We worked with persons of all ages, including entire families. Over a course of several weeks we dealt with approximately several hundred people. A majority of people were from distant locations, and without direct daily contact with the sea. Many different issues were raised by the participants during our activities. The issues depended mostly on the age and place of residence of the participants. We learned that the weather was an important factor, since it moderated numbers of participants and their level of interest.

Another aspect of the first stage of the I Live by the Sea Project involved an art gallery. The Sopot Science Association, together with Today We Have and the Sea for Society project team created a gallery of marine related photographs made by researchers from the Institute of Oceanology Polish Academy of Sciences. These photographs had been made during their research cruises in the Baltic Sea and the Arctic. From over 70 photographs a total of 33 were selected and printed in 50 × 70 cm size. Each of the photos received a title and a short story.

Then we organized a number of public exhibitions, from a pedestrian area in the sea resort of Sopot (2015), through the Ministry of Environment in Warsaw (2015), to the Science Day (2017). We had official openings of the exhibitions, with authors invited, which provided a great platform for discussions. Just like in other cases we had both adults and children, including families as visitors. Beside the evident beauty of the photographs, the visitors had an opportunity to learn and discuss marine issues with the researchers who made the photographs. Thus, in this case, we combined artistic aspects with a professional approach to non-formally educate about research in the Baltic and the Arctic.

3.2.2 I Live by the Sea. International Contest (http://todaywehave.com/CONTEST.html)

The I Live by the Sea International Youth Contest is a continuation of a national project which took place in 2015 and early 2016 in Poland. Our goal is to spread and share knowledge about marine environments and marine protection issues as well as the exchange of information among users of different seas and marine oriented regions in a non-formal approach. The I Live by the Sea International Youth Contest also aims to involve youth of different natural and cultural heritage to share their views on marine issues in a creative manner.

The idea behind this project involves a worldwide contest for young persons up to the age of 18. The 2017 edition of the contest was held in three age categories, 5–12, 13–15 and 16–18.

Participants were required to send a photo of "their sea" along with a short text. This way, not only the photo was taken into consideration but also a relevant description of a region, a special feature of the place, an explanation of the depicted problem, etc.

This project is based on the previous experience of a three-stage contest for young learners which has been described in Sect. 3.2.1 above. Just like in the previous national editions, the 2017 edition aimed at spreading and sharing knowledge about marine environments and marine protection issues as well as the exchange of information among users of different seas and marine oriented regions. Through their participation we managed to involve youth of different natural and cultural heritage to share their views on marine issues in a creative way. We obtained 136 submissions from 12 countries, including Trinidad and Tobago, USA, Poland, the Philippines, Romania, Kazakhstan, the Republic of Mauritius, Spain, Cambodia, Turkmenistan, the Republic of Mozambique and the Sultanate of Oman.

In the final stage of the 2017 I Live by the Sea edition, a photo exhibition was presented in the United Nations Headquarters in New York City on 6 June 2017, as a side event of the United Nations Oceans Conference (5–9 June 2017). We had hundreds of visitors from over a hundred countries during this event.

4 Summary and Conclusions

Our world is becoming continuously and increasingly "smaller" and more competitive. Research and development reach beyond borders which results in new opportunities, but also requires societies to face more complex challenges. In order to deal with these challenges, modern citizens will have to be equipped with a better understanding of science and technology, in order to be able to participate actively and responsibly in science-informed decision-making in a knowledge-based world. This means that all societal actors, from user groups, specialists and stakeholders, through professionals, to enterprises or the industry play crucial roles.

The modern world is short of science-knowledgeable citizens at all levels of society. Even though, over the last several decades, an increasing number of graduates in science has been observed, there is no increase in the number of persons keen on pursuing science careers.

We have to realize that science education, research, innovation and practices must become more adapted to the requirements and ambitions of a modern society and reflect its values. They should provide society with the science that they need and support citizens of all ages and talents in creating positive attitudes to scientific endeavors. We must look for new and more innovative means to sustain natural curiosity and cognitive resources of young learners. We need to support the educational process which will ensure that future researchers and other actors will be equipped with modern knowledge, motivation and a sense of societal responsibility to participate actively in the process of further innovation. Some of the examples have been presented in this paper and in a synthetic form in (Table 2).

In order to efficiently collaborate and exchange information and knowledge of and about science and science communication, as well as to find solutions for global challenges of the global society, we should actively co-operate at an international level.

Overall, the examples of non-formal education activities which have been described in this paper provided all participants, including our teams, with a mutually valuable experience, which provided important contribution for future

Table 2 Summary of the activities described in the paper

Project title	Target group	Range of activities	Estimated numbers of participants	Description of project results
Sea for Society	The entire society, from young learners, through the general public, professionals engaged in marine issues to decision makers	Regional	Hundreds to thousands	Results show increased enjoyment of science lessons and extra-curricular activities, increased confidence and understanding in learning science, development of transferable and practical skills, increased awareness of the importance of science to society
I Live by the Sea. Stages I–III	From young learners of all ages to the general public	Regional to national	Hundreds	Involvement of various groups of citizens from young learners to adults in learning about marine issues through art and through active creation of art related to marine issues
I Live by the Sea. International Contest	Young learners of ages 5–18	Global	Hundreds to thousands	Global impact of the Project, over a hundred participants and thousands of viewers from all over the world learning/teaching about regional and global marine issues

collaboration with schools and life marine educational programs. In particular, our positive experiences relate to:

- The ability to discuss marine and environmental problems from different perspectives;
- The use of visual arts as an innovative approach to teaching about the environment through mobilization of youth to realize various tasks which lead to the presentation of gained knowledge;
- Youth involved in activities within and outside of the classroom;
- Through involvement of parents and friends, we managed to reach the broader public, including adults;
- The many great ideas our participants came up with, e.g. making films on the beach or creating plots for theater plays.

Acknowledgements We hereby thank Joanna Piwowarczyk from the Institute of Oceanology Polish Academy of Sciences for her valuable input and we acknowledge the Sea for Society Program for providing useful data.

References

Ainley M, Ainley J (2011) A cultural perspective on the structure of student interest in science. Int J Sci Educ 33(1):51–71

Anderhag P, Emanuelsson P, Wickman PO, Hamza KM (2013) Students' choice of post-compulsory science: in search of schools that compensate for the socio-economic background of their students. Int J Sci Educ 35(18):3141–3160

Beaumont NJ, Austen MC, Atkins JP, Burdon D, Degraer S, Dentinho TP, Derous S, Holm P, Horton T, van Ierland E, Marboe AH, Starkey DJ, Townsend M, Zarzycki T (2007) Identification, definition and quantification of goods and services provided by marine biodiversity: implications for the ecosystem approach. Mar Pollut Bull 54(3):253–265

Brander L, Baulcomb C, van der Lelij JAC, Eppink F, McVittie A, Nijsten L, van Beukering P (2015) The benefits to people of expanding Marine Protected Areas. VU University, Amsterdam, The Netherlands

Bray B, France B, Gilbert JK (2012) Identifying the essential elements of effective science communication: what do the experts say? Int J Sci Educ Part B Commun Public Engagem 2 (1):23–41. https://doi.org/10.1080/21548455.2011.611627

Costanza R (1999) The ecological, economic and social importance of the oceans. Ecol Econ 31 (2):199–213

Costanza R, D'Arge R, de Groot R, Farber S, Grasso M, Hannon B, Limburg K, Naem S, O'Neil RV, Paruelo J, Raskin RG, Sutton P, van den Belt M (1997) The value of the world's ecosystem services and natural capital. Nature 387(6630):253–260

Domegan C, McHugh P, Devaney M, Duane S, Hogan M, Broome B, Layton R, Joyce J, Mazzonetto M, Piwowarczyk J (2016) Systems-thinking social marketing: conceptual extensions and empirical investigations. J Mark Manage 32–2016(11–12):1123–1144

European Commission (2015) Science education for responsible citizenship

George JM, Lubben F (2002) Facilitating teachers' professional growth through their involvement in creating context-based materials in science. Int J Educ Dev 22(6):659–672

http://seaforsociety.eu

http://www.oceansanctuaryalliance.org/

http://www.un.org/sustainabledevelopment/

Kotynska-Zielinska I, Piwowarczyk J, Zielinski T (2016) School children performing on marine issues related to the Baltic Sea and the Arctic. In: Poster 12, CommOcean Conference, Bruges, 2016

Kuhn D (1993) Science as argument: implications for teaching and learning scientific thinking. Sci Educ 77(3):319–337

Lubchenco J, Grorurd-Colvert K (2015) Making waves: the science and politics of ocean protection. Science 350:382–383

Ronnback P, Kautsky N, Pihl L, Troell M, Soderqvist T, Wennhage H (2007) Ecosystem goods and services from Swedish coastal habitats: identification, valuation and implications of ecosystem shifts. Ambio 36(7):534–544

Stocklmayer SM, Bryant C (2012) Science and the public—what should people know? Int J Sci Educ Part B Commun Public Engagem 2(1):81–101. https://doi.org/10.1080/09500693.2010.543186

The Convention on Biological Diversity's High-Level Panel on Global Assessment of Resources for Implementing the Strategic Plan for Biodiversity 2011–2020 (see https://www.cbd.int/financial/hlp.shtml)

Wallmo K, Edwards S (2008) Estimating non-market values of marine protected areas: a latent class modeling approach. Mar Resour Econ 23(3):301–323

Weslawski JM, Andrulewicz E, Kotwicki L, Kuzebski E, Lewandowski A, Linkowski T, Massel SR, Musielak S, Olańczuk-Neyman K, Pempkowiak J, Piekarek-Jankowska H, Radziejewska T, Różyński G, Sagan I, Skóra KE, Szefler K, Urbański J, Witek Z, Wołowicz M, Zachowicz J, Zarzycki T (2006) Basis for a valuation of the Polish exclusive economic zone of the Baltic Sea: rationale and quest for tools. Oceanologia 48(1):145–167

Big Data Analysis of the Environmental Protection Awareness on the Internet: A Case Study

Dorota Majewicz and Jacek Maślankowski

Abstract Big Data tools allow us to make an exhaustive analysis of specific types of information within all the specified websites on the Internet. They can provide large quantities of information on the current discussion on specific topics. In the paper, we have decided to provide an analysis of selected websites on environmental protection with the choice of three types of websites and three different countries in order to be able to provide a more diversified perspective to examine the most commonly discussed topics that appear there.

Keywords Big data · Environmental protection · Unstructured data
Data analysis · Data quality

1 Introduction

In recent years the role of Big Data tools in environmental protection has been increasing. In particular, there are theories and methods that can be used to evaluate environmental performance (Song et al. 2016), and such applications can be used by the government sector (Environmental Law Institute 2014; Kim et al. 2014). Ecologists see big data as a promising data source (Hampton et al. 2013). The rise of big data applications related to environmental protection can be the result of the fact that, according to some researchers, we live in the world of the Internet of Things (Lu et al. 2015; Li et al. 2011) and in the near future, the Internet of Everything. Smart meters are used to measure energy consumption (Keeso 2014) or pollution in the air (Weranga et al. 2014). For many years now, we have been able to observe a correlation between energy consumption and air pollution (Duncan et al. 1975). Regarding smart meters, there exist air-quality-monitoring-stations in

D. Majewicz (✉) · J. Maślankowski
University of Gdansk, Gdansk, Poland
e-mail: dorota.majewicz@ug.edu.pl

J. Maślankowski
e-mail: jacek.maslankowski@ug.edu.pl

T. Zielinski et al. (eds.), *Interdisciplinary Approaches for Sustainable Development Goals*, GeoPlanet: Earth and Planetary Sciences,
https://doi.org/10.1007/978-3-319-71788-3_4

cities, however their number is not significant at present (Zheng et al. 2013). For example, in Poland, in the Tri-City subregion with ca. 1 million inhabitants, the data published are gathered from nine stations (aqicn.org 2017). Regarding environmental protection, there are also data available on hydrology systems to monitor a potential risk of flooding (Vitolo et al. 2015). In the United States, the data are processed and published to calculate the value of EQI—Environmental Quality Index—that uses air, water, land, built and socio demographic domains transformed variables (USEPA 2017). Because there are numerous examples of such types of research in the literature, we have decided to conduct an analysis of people's awareness of environmental protection that can be observed on the Internet.

The goal of this paper was to ascertain how and to what extent people are aware of environment protection and how websites are supporting such activities. For the purpose of this paper, we decided to make a partial analysis of selected websites as well as Internet forums. The results of our analysis are presented herein below.

This paper has been divided into five parts: after the introductory section, we go on to discuss the methodology used in order to make the analysis in the second part of the paper, where we concentrate on big data as the main set of tools used there. After the part where we describe the methodology used, we provide information on current websites on environmental protection that have been used for the analysis, and we eventually show a case study of the analysis with a discussion on the current use of Internet forums in the fourth part. The closing one presents conclusions resulting from the case study.

2 Methodology

In the paper we have used Big Data tools supported by Web Mining methods. The goal of Web Mining is to discover useful information or knowledge from websites (Liu 2011), and, for the purposes of our research, we have been concentrating on unstructured data. Although such data do not necessarily need to be textual (e.g., images, videos) (Inmon and Nesavich 2007) we have been using textual data exclusively. In fact, such data are human-sourced information available on website content (Jain 2017).

Our work focuses on the issue of finding useful information on websites with regard to current research conducted in the field of increasing people's awareness on environmental protection and, additionally, various topic-related activities to achieve it. To accomplish the task we decided to select topic-oriented websites on environmental protection (more details have been presented in the third part of this paper). We have prepared a set of tools to find the most current topics discussed on such websites. In our attempt to find such topics, we have been trying to find whether subject-oriented websites are more concerned with reducing air pollution, saving natural environment or other issues, not bearing in mind the correlation between them. Our partial goal has been to show how we can use Big Data tools to obtain answers to the questions mentioned above.

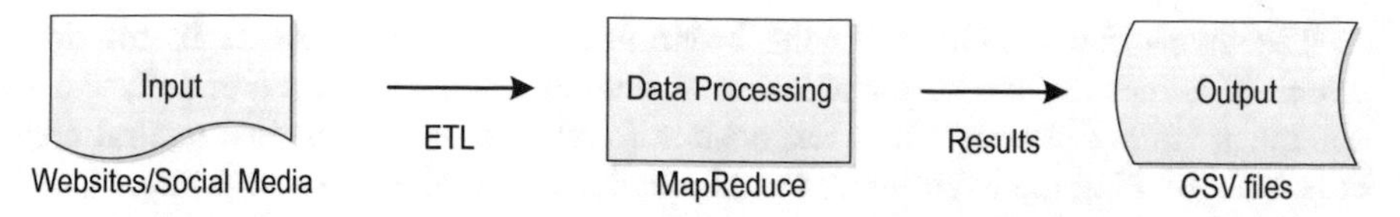

Fig. 1 Procedure of data collection and processing

In order to accomplish the main goal of the paper, we decided to use the procedure of data collection and analysis presented in Fig. 1.

We have collected the data from websites using the ETL (Extraction, Transformation, Loading) process. The extraction phase has been done in the memory as well as the transformation phase. The results have been stored in memory and processed by MapReduce algorithms, mostly word count and regular expressions. Next, the results of the analysis have been stored in the CSV file for further analysis and visualization. Our choice of the software used has been Apache Hadoop and Spark environment, while Python has been the primary programming language used in the scripts.

The algorithms used to make a word count analysis were prepared according to natural language processing rules, regarding lemmatization, stemming, stop words, and other techniques that increase the reliability of results. From the technical point of view, all the libraries as well as the data used in this analysis are freely available, and this tool can be used for further analysis of other environmental protection-oriented websites and social media channels.

3 Environmental Protection

We decided to limit our choice of websites for the analysis to three different types, namely government official websites, companies/laboratories that emphasize the need for environment protection as part of their official policy, and, lastly, non-governmental organizations that are involved in environmental management, advocacy and conservation methods in order to decipher whether the information provided and the message conveyed is coherent within one organization. We analyzed the data in three different countries, namely Poland, Switzerland and the USA, to ascertain to what extent the message about environment protection is consistent among them. Our choice of the countries is justified by the fact that both Poland and Switzerland are currently concerned about the wood resource policy, preservation of national parks and biodiversity, and the challenges and discussions appear diverse compared the two government websites. The choice of the US Environmental Protection Agency was random, with an emphasis put on its exhaustive contents and unequivocal engagement in the promotion of environmental awareness among its citizens.

The Swiss Federal Office for the Environment[1] (FOEN) presents its mission, which it defines as "the sustainable use of natural resources including soil, water, air, quietness and forests." It is responsible for the protection against natural hazards, safeguarding the environment and human health against excessive impact, and conserving biodiversity and landscape quality. It also ensures responsibility for international environmental policy.

Additionally, the Swiss Federal Department of Foreign Affairs presents a website on the main issues of concern to the environment in a part entitled "Key aspects of environmental protection in Switzerland.[2]" Natural water resources are emphasized, and the main areas of concern for the Swiss government are the protection of natural resources (with an emphasis put on the so-called "green economy"); higher density urban planning, with two main objectives being a more economical use of available space and the limitation of excessive spread of building zones; the fight against global warming, with the focus on the reduction of the emissions of greenhouse gases and justification for carbon tax rise; the responsibility of the preservation of water quality; a special strategy for maintaining biodiversity; a policy on imposing strict standards of CO_2 emissions on the industry to improve air quality; soil protection with support for farmers; and, lastly, the importance of conservation of the countryside as a priority over other plans, including those concerning energy, industry or transport. A recently discussed issue of the Wood Resource Policy has been briefly presented as a one "to ensure that wood from Swiss forests is supplied, processed and used in a way that is sustainable and resource-efficient. In doing this, it makes a major contribution to forest, climate and energy policy." In order to promote and explain current trends and challenges, the government issues a magazine entitled "Environment," which is published quarterly online in German, French and in Italian, and "each issue focuses on a specific theme, presenting the latest knowledge in the field and providing keys to acting more sustainably." Similarly to Poland, Swiss national parks are important, and "they were created to promote the distinctive scenery—both natural and cultivated—with the biodiversity typical of the local region and an environment in which people can find inspiration over and again, whether from the local culture, economy or culinary traditions." The description is followed by an encouraging message: "[P]arks house a myriad of small wonders, waiting to be discovered, experienced, heard and savoured." Thus, the organization of the government websites, a coherent explanation of the need for environmental awareness, and, additionally, a clear and concise explanation for the necessity of preservation and/or protection of selected areas serves as an encouragement for such activities and promotes an environmentally friendly community. The overall message conveys a friendly atmosphere that supposedly aims at encouraging people to share an ecologically friendly approach.

[1]https://www.bafu.admin.ch/bafu/en/home.html.

[2]http://houseofswitzerland.org/swissstories/environment/key-aspects-environmental-protection-switzerland.

The Polish Ministry of Environment[3] describes its mission as follows: "through its input into national policies, [the Ministry] fosters the environment both domestically and globally, and ensures the long-term, sustainable national development with respect to natural heritage and human rights to meet the needs of both the present and the future generations," while in the second passage called Vision, there is a statement saying that "[The] Ministry of the Environment, as the state-of-the-art, professional institution which enjoys social trust, provides for rational management of natural resources and environmental education of the general public, and is open to cooperate in the field of the environment." The main page provides readers with the latest news related to environmental issues, information on the Ministry with names and contact details, and additionally a separate tab for an issue that has recently caused heated debates in Polish society, namely the large-scale logging in the Białowieża Forest. The page presents "the infographic illustrating the situation in the Białowieża Forest with regard to the research and monitoring experiment carried out at its area," but it lacks a description for readers who are interested in the topic. The description itself begins with an emotional message, which is unusual for official government-related websites: "Many people talk about it although they know little about it. They have never been to the Białowieża Forest. They base their judgment on information available in the media, which is often created by people who do not know much about this unique forest." Readers are encouraged to "find out about the Białowieża Forest in order to discover its true nature," which adds little to provide a concise explanation of the situation based on the results of research conducted in the area that would objectively relate to discussions that have been addressed by the public. As a result, the information provided lacks a straightforward account of the situation or future plans for perseverance of the natural resources, that is the national park; the website states, "We shall see how the story unfolds." The website claims that "the research is a starting point. It is a base on which we shall construct a system that will make it possible for all those interested to track and comment on the changes taking place in the ecosystems of the Białowieża Forest," and citizens are encouraged to follow the website: "The internet application that we are planning to develop will let everyone observe the changes taking place personally. The whole world will be able to make use of the knowledge that we will make available and find out for themselves in what way the centuries old heritage of mankind helped preserve this magnificent forest—the Białowieża Forest. We encourage everyone to be a part of this fascinating scientific adventure." However, we assume that encouraging citizens to have a more active attitude would help them benefit more with regard to the issue of protection and preservation of natural parks. There is no information available on the habits of forest's inhabitants or advice on appropriate conduct in a protected natural area (such a case will be presented on the following passages concerning the EPA). However, there is an attempt to involve a larger community in the discussion about the forest by sharing its monitored parts online: "We see a tight network of

[3]https://www.mos.gov.pl/en/ministry/.

monitored areas, where the research was carried out. Each of the 1400 points is an area of 400 m squared, indicated by circles having a radius of 11.28 m. Each of these points will be available for observation, it will be possible to find out the species inhabiting that particular area, as well as compare various parts of the forest and personally assess how human activity, or its absence, impacts the ecosystem of the Forest. Pictures and films will be available. Everyone will be able to comment on what they see. This will be a social action on a global scale. An invitation to the real world of the Białowieża Forest by means of a virtual tool".

The English website of the Polish Chief Inspectorate of Environmental Protection is only partly translated (some articles or pieces of information are presented in Polish, regardless of the English version) and it is divided into four main parts: About the Office (with names of the Chief and Deputy Chief Inspectors of Environmental Protection and the General Director of the Chief Inspectorate of Environmental Protection), Waste Management (with two links regarding "Methodological guidelines on treating damaged vehicles as waste in the trans-boundary shipment of waste"), Major Accidents (with seven legal acts on the prevention of a variety of accidents), and State of the Environment (with the presentation of "the State Environmental Monitoring Programme" and "State of Environment Reports[4]"). The government policy on major issues of environment protection are not present on the website; however, there is more emphasis on the cooperation with the European Environment Agency and legal acts; the reader is redirected to external links if he/she wishes to obtain more information, as in the case of Poland's involvement in SOER 2015 (The European environment: The state and outlook 2015 report). In comparison to the Swiss website, the Polish one provides the visitor with far less information on the environment policy of the government, with little emphasis on the need for its protection, and much important and valuable information in the English version of the website is available only in Polish (e.g., workshops, conferences, and other initiatives). An overall impression of the Polish government's websites on environment preservation and protection is that they do not make an important contribution to encouraging a conscious ecologically-friendly lifestyle, as there is little information that might be useful in promoting it.

To contrast the contents discussed above, we have compared it with the official website of the United States Environmental Protection Agency[5] (EPA or sometimes USEPA), whose mission is to protect human health and the environment, i.e., air, water, and land, by writing and enforcing regulations based on laws passed by the Congress. Similarly to the Swiss government's website, the US website, apart from "About EPA" and "Laws and Regulations" parts, presents an extensive unit on environment-related topics, which is divided into thirteen diverse sub-sections (e.g., Chemicals and Toxics, Greener Living, or Land, Waste, and Cleanup) and which provides citizens with information on hazardous substances, actions taken in order

[4]http://www.gios.gov.pl/en/.

[5]https://www.epa.gov/home/forms/contact-epa.

to improve the environment, and instructions on possible ways to improve and protect the environment.

Once the website is visited, a message with a request to leave feedback designed to measure the entire experience, appears on the screen. The fact that a person is asked for help in order to improve the website created by a government agency serves as the first step toward active participation in the process of creating a website for citizens as a unique community. The topics on the website are organized alphabetically, which suggests that their number is significant, and they vary from information on seemingly insignificant issues such as bed bugs, to monitoring air emissions, land and waste management, pest management, or recycling electronics, to mention but a few. The content and the number of issues addressed at the website imply deep concern for the environment and the consciousness of the fact that citizens should be sufficiently informed and encouraged to ask questions on problems that might seem to be insignificant. The language of the informative passages, similarly to the one on the Swiss websites, is rather simple; the style is objective, informative and encourages the reader to participate in the exchange of information. This is similar to the message and an overall impression of several non-government organizations in all the three countries, where the need for sustained development with an emphasis put on environment protection is emphasized. The examples of such organizations, which, in the opinion of the present authors, are relevant to those created by governments, are—among many—Pro Natura in Switzerland: https://www.pronatura.ch/79, Agro-Group in Poland http://www.agro-group.org/eng/, and Earth Policy Institute in the USA http://www.earth-policy.org.

4 Case Study and Discussion

Our presumption in the paper was that both government and non-government organizations whose work is involved with issues of environmental protection may have a major impact on people's awareness of the fact that actually every person is responsible for protecting the environment. Owing to this fact, we have concentrated on social media channels as well as Internet forums, where people can read and discuss particular topics related to environmental protection, its problems and possible solutions.

In the case study we decided to show two different examples—the EPA from United States and the Ministry of the Environment of the Republic of Poland. This decision was motivated by the possible influence on society in terms of people's awareness of the need for environmental protection despite numerous challenges. Thus, legal authorities should derive reliable information on the current situation regarding environmental protection. Our goal was to find if there is an insufficiency of topics and discussions on websites (as mentioned in Sect. 3) or social media channels described in this section.

Below, we have provided examples of the analysis with Big Data tools regarding topics discussed on particular media channels. Firstly, we decided to examine how

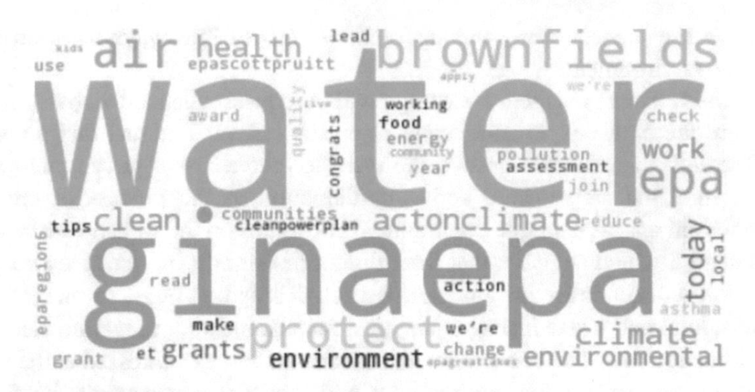

Fig. 2 The most popular topics on the EPA's social media channel

the EPA (The US Environmental Protection Agency) uses its social media channels to increase citizens' awareness of environmental protection. We decided to conduct an analysis of this with the Big Data tools to acquire and analyze tweets on the EPA's official Twitter account. As of August 2017, this social media channel is followed by 569 thousand Twitter users. Since the beginning of this channel, ca. 13.8 thousand tweets have been posted or retweeted. We decided to use a Word cloud to make a visualization of the topics most published on EPA's Twitter account, and we used a word count analysis to provide the results (Fig. 2).

The results of our study show that the most popular terms are related to clean water, grants, food, energy, pollution, and climate. It is connected with the potential issues regarding the environment and its protection of which people must be aware. More detailed topics have been discussed later in the forthcoming section of the paper.

The second social media channel that was analyzed was the official Twitter profile of the Ministry of Environmental Protection of the Republic of Poland. This account is observed by 5.6 thousand Twitter users. It has had 4.6 thousand tweets, as of August 2017. The language used on this social media channel is Polish, thus we had to change an algorithm to fulfill all issues related to lexical forms of this language. The results are presented in Fig. 3.

The results of the word-frequency analysis allow us to formulate the conclusion that this account is used mostly to educate people and explain how to protect the environment. In particular, the most relevant topics for the society are discussed and tweeted numerous times. For example, there is a large quantity of information on the Białowieża Forest (in Polish: *Puszcza Białowieska*) and logging (in Polish: *wycinka drzew*). It means that the main role of this social media channel is to discuss current issues related to environmental protection in Poland.

The main conclusion which arises from our analysis is that the organizations analyzed support environmental protection by publishing recent news related to the issue. Below are examples of tweets from the EPA account that encourage people to save the environment:

Fig. 3 The most popular topics on the Polish Ministry of the Environment's social media channel

(1) *Did you know that flooding can cause harmful mold to grow? Stop the mold!*
(2) *Nutrient pollution is 1 of U.S.'s most widespread, costly & challenging environmental issues.*

On the other hand, there are also posts that help to protect people from the harmful effects of natural environment pollution or ozone zone:

(1) *Exposure to #radon indoors causes about 21,000 lung cancer deaths each year. Protect yourself.*
(2) *It's summertime! That means the days can get hot. Wear sunscreen and plan for the heat.*

The Polish Ministry of the Environment is using its social media channel to provide instant information on the initiatives that are intended to save the environment, e.g., funds for environmental protection, garbage collection opportunities, and various types of success of different regions in the protection of the environment, e.g.:

(1) *The city Leszno has achieved 34.7% of selective garbage collection. It is more than the average value for Poland.*
(2) *Success—1.3 billion EURO for investments on garbage sector.*

Having conducted a deep analysis of the topics covered on a number of websites and social media channels, we can argue that people's awareness of environmental protection can be increased by subscribing to such social media channels or related discussion lists. However, there is still lack of quick information concerning suggestions and solutions to actions that can help ordinary people to become more environmentally friendly.

We suggest that it would be more expedient to publish more topics on environmental protection in terms of what people can do to save the environment. Such topics are present on websites of environmental organizations, but they are not repeated or retweeted many times and may disappear in the large amount of information provided on those channels. That is why we recommend that public authorities exercise their power and influence to educate people more frequently on the basic activities that may lead to changes such as a decrease in air pollution. It will help to be in tune with the current trends on environmental protection, especially taking account of the fact that almost every month new forms of environmental protection are being developed by researchers.

5 Conclusions

This analysis of diverse types of organizations and the information provided and discussed on their websites, in addition to the choice of three different countries, allows us to reach several conclusions related to the issue regarding people's environmental protection awareness.

In the paper we discussed the impact that websites and social media channels of government and non-government authorities may have on creation and providing information as well as expanding knowledge among citizens to help save our planet. Our analysis has led us to formulate assumptions that the increase in the number of posts with easy steps and activities for people to follow in order to be more eco-friendly will extend their awareness of the necessity to protect the environment. A local view of the problem will have a global effect, e.g., to use a bicycle rather than a car whenever the latter is not necessary, to increase the green areas in gardens or to save water, which, as a result, will help people gain a broader perspective on a particular issue related to the environmentally friendly lifestyle and community behavior.

Internet communication has provided people with a chance to reach the global community instantly and give Internet users brief and up-to-date information. Decisions of both government and non-government officials to expand their activities online and encourage people to subscribe to their Twitter or Facebook accounts not only gives them information on their followers and, as such, helps them to personalize their social media contents with regard to the message style, but it also provides them with information that is helpful in making a respondent-tailored choice of activities in the future.

References

Aqicn.org, Data from monitoring stations for Tri-city subregion, http://aqicn.org/map/poland/gdansk/sopot/pl/. Accessed 2 Aug 2017

Duncan JR, Noll KE, Davis WT (1975) air pollution control and industrial energy production. Ann Arbor Science Publishers

Environmental Law Institute: Big Data and Environmental Protection: An Initial Survey of Public and Private Initiatives (2014) https://www.eli.org/sites/default/files/eli-pubs/big-data-and-environmental-protection.pdf. Accessed 2 Aug 2017

Hampton SE, Strasser CA, Tewksbury JJ, Gram WK, Budden AE, Batcheller AL, Duke CS, Porter JH (2013) Big data and the future of ecology. Front Ecol Environ 11(3). doi:https://doi.org/10.1890/120103

Inmon B, Nesavich A (2007) Tapping into unstructured data: integrating unstructured data and textual analytics into business intelligence. Prentice Hall

Jain VK (2017) Big Data and Hadoop. Khanna Book Publishing

Keeso A (2014) Big data and environmental sustainability: a conversation starter. Smith School Working Paper Series. http://www.smithschool.ox.ac.uk/library/working-papers/workingpaper%2014-04.pdf. Accessed 2 Aug 2017

Kim GH, Trimi S, Chung JH (2014) Big-data applications in the government sector. Commun ACM 57(3):78–85. https://doi.org/10.1145/2500873

Li S, Wang H, Xu T, Zhou G, Yang D (2011) Application study on internet of things in environment protection field. Informatics in control, automation and robotics, vol 2. Springer, pp 99–106. doi:https://doi.org/10.1007/978-3-642-25992-0_13

Liu B (2011) Web data mining. Exploring hyperlinks, contents and usage data, 2nd edn. Springer

Lu SQ, Xie G, Chen Z, Han X (2015) The management of application of big data in internet of thing in environmental protection in China. In: 2015 IEEE first international conference on big data computing service and applications, Redwood City, CA, pp 218–222. doi:https://doi.org/10.1109/BigDataService.2015.68

Song M-L, Fisher R, Wang J-L, Cui L-B (2016) Environmental performance evaluation with big data: theories and methods. Ann Oper Res. https://doi.org/10.1007/s10479-016-2158-8

USEPA (2017) https://catalog.data.gov/dataset/usepa-environmental-quality-index-eqi-air-water-land-built-and-sociodemographic-domains-transf. Accessed 3 Aug 2017

Vitolo C, Elkhatib Y, Reusser D, Macleod CJA, Buytaert W (2015) Web technologies for environmental Big Data. Environ Model Softw 63:185–198. https://doi.org/10.1016/j.envsoft.2014.10.007

Weranga KSK, Kumarawadu S, Chandima DP (2014) Smart metering design and applications. Springer

Zheng Y, Liu F, Hsieh H-P (2013) U-Air: when urban air quality inference meets big data. In Dhillon IS, Koren Y, Ghani R, Senator TE, Bradley P, Parekh R, He J, Grossman RL, Uthurusamy R (eds) Proceedings of the 19th ACM SIGKDD international conference on knowledge discovery and data mining (KDD'13), ACM, New York, NY, USA, pp 1436–1444. doi:https://doi.org/10.1145/2487575.2488188

Pluralistic Forms of Ensuring the Social and Democratic Balance as Part of Sustainable Development

Bartosz Maciej Wiśniewski

Abstract This paper was inspired by questions asked by John Elkington, the Environment Foundation Chair who wanted to know if it is possible for a democratic world to vote ourselves into sustainable future. The author analyzes the state of global democracy and economy in a context of pluralistic "horizontal" theory of democracy by Robert Dahl in order to answer the questions presented by the Chair of Environment Foundation. This article focuses on the importance of pluralistic forms of ensuring the social and democratic balance.

Keywords Pluralistic theory of democracy · Democratic balance
Sustainable development · Polyarchal democracy

1 Introduction

Sustainable development is not only a matter of environment and economy. It is not only about human population rising too high and most certainly it is not only about renewable energy versus fossil fuel. It is about all those things and much more. Public participation and access to information have long been recognised as a problem in the fight for sustainable development. Today one cannot talk about sustainability of growth without addressing the problem of the relationship between democracy and sustainable development. The scale and pace of environmental and social change create an urgent need for action. The questions asked by John Elkington the Environment Foundation Chair, seem to sum up the problem perfectly:

> Can we vote our way to a sustainable future for a world of 9 billion people - or are new forms of leadership going to be necessary? What kinds of systems of governance and decision-making best place countries and people to move towards more sustainable forms

B. M. Wiśniewski (✉)
Ateneum-University in Gdansk, Gdansk, Poland
e-mail: b.wisniewski@ateneum.edu.pl

T. Zielinski et al. (eds.), *Interdisciplinary Approaches for Sustainable Development Goals*, GeoPlanet: Earth and Planetary Sciences,
https://doi.org/10.1007/978-3-319-71788-3_5

> of development? Can the time-frames of democratically elected governments deliver sustainable development? If not, what needs to be done - and by whom?
>
> (The Environment Foundation 2017)

The Environment Foundation began work on Democracy and Sustainable Development in 2007 and it still to this day continues its work with different partners to make sure that democracy delivers sustainable development (ibidem). This paper aims to give the policy makers and organisations like the Environment Foundation a broad analysis of pluralistic forms of democracy to raise awareness of existing tensions and good practices. Only practical actions based on a strong democratic model will be able to support decision-making fit for the challenges of twenty-first century and sustainable development.

The democratic theory that the author of this paper believes to be the most suitable for ensuring a sustainable social and democratic balance as part of sustainable development is the model proposed by Robert Dahl. Using a spatial metaphor, his model can be described as a "horizontal" theory of democracy. It contradicts the earlier "vertical" theory of democracy in which we have a highly unsophisticated sovereign who, mainly because of their intellectual limitations, must cede their decision-making powers upwards to the "elite". This so-called "elite" is given the power to make political decisions on behalf of the sovereign. In this theory there is virtually no communication between the society and the policy makers. There are no institutions, organizations and intermediaries, only cyclical elections, deciding which "elite" group will rule. Dahl decided that this model of democracy was oversimplified. This simplification was related mainly to the existence of intermediaries serving as a communication bridge between these two levels. Those broadly understood intermediaries are in R. Dahls' theory the wealth of organized civil society. Those intermediaries can be anything between different associations, clubs, trade unions and business organizations (Dahl 1956). In truth, Dahl himself does not call this model a horizontal one, but rather describes it as polyarchal democracy but for the sake of this paper author will stick to the more easily understandable nomenclature.

2 Sustainable Development and Democracy

When thinking about sustainable development we have to realize that it can only take place in a political system that itself is sustainable. The "vertical" model of democracy has proven time and time again to not be sustainable in a long run. As an example of failings of "vertical" model we can look at the political situation in Poland after the 2015 elections. The political situation in Poland after the 2015 elections has been rated by a large part of the population as a threat to democracy. The actions of the government can be characterized by gross incompetence, consistent destruction of the achievements and international image of the country. Nepotism affected the interests of many social groups and at the same time the image of political order (Tłokiński and Wiśniewski 2017). There was a visible need for communication between the policy makers and the rest of society. But without strong intermediaries in the form of associations, clubs, trade unions and business organizations there was no way for the sovereign to voice its dismay. We witnessed the birth of the Committee for the Defence of Democracy. It is a social structure that aggregates diverse groups of interests that may be usually represented by mentioned before different associations, clubs, trade unions or business organizations (ibidem). That example shows us that in democracy there is a strong need for dialog in times other than election and in pluralistic theory of democracy by Robert Dahl that communication is supposed to take place through aggregated groups of interest like the Committee for the Defence of Democracy.

In pluralistic theory of democracy, active citizenship participation is a key element that guarantees sustainability of political system. The main ingredients described by Robert Dahl as crucial when characterizing a healthy and sustainable democratic system are:

1. multiplicity of sources of information;
2. freedom of expression;
3. freedom of association;
4. universal electoral law (Dahl 1972).

Considering the fact that we are living in an age of Internet, the multiplicity of sources of information in non authoritarian states is a given. But what about the three remaining ingredients? When describing democracy is hard to imagine it not having all of those attributes. The problem seems to be not with the lack of those ingredients but with the willingness of democratic society to exercise those rights. Without the social will to take part in the democratic process there is no way of ensuring the democratic balance of power. That is why it is so important to educate the society about the importance of active citizenship participation. Education is also crucial in spreading the awareness of humanity's survival being interlocked with the state of our environment (Laneve 2017).

The big problem with sustainable development in democratic states of the First World seems to be the constant pressuring of the Third World Countries to do the limiting: reduce population, preserve rain forests, conserve water, etc. It is not easy, then, to reflect on sustainability from the perspective of the urban poor who make up a large part of Third World Countries population and for whom many basic needs remain unmet. There is in fact some discomfort with the topic, as 'sustainability' is in many ways an induced concern in urban poor communities who are struggling to satisfy the basic material needs essential to their own personal and social development (Joseph 2001). That is why it is of utmost importance to create such a model of democracy in which the responsibility of maintaining sustainable development is not shifted primarily unto the less developed countries. Furthermore, it is crucial that within the above-mentioned model we will have a society that understands the need for self regulation. Not only that but the same society must be ready to implement that self regulation in real life situations.

One might ask how to achieve such a democratic society that will be willingly sacrificing its privileges? The answer to that question is through a dialog between policy makers (the so-called elite) and society (sovereign). As mentioned before, that kind of dialog cannot take place just in brief times of elections. If the public is not made a part of that dialog, they will not be able to understand and to accept the need to adjust their lifestyle. This is where the pluralistic forms of ensuring the social and democratic balance come into play.

The above-mentioned social dialog must take place through aggregated groups of interests. That mechanism of aggregation of the interests should take place at the level of the different social classes (Bartolini and Mair 1990) to ensure transparency and social acceptance. Whether or not those aggregated groups of interests remain in permanent opposition to each other is a real problem for the policy makers who will have the responsibility of deciding which of the interests are more important for maintaining a sustainable growth. Going back to the questions asked by John Elkington that the author quoted at the beginning of this paper it is important to analyze the state of world democracy.

3 Analysis of Global Democracy

In this paper, the author will carry out the analysis of global democracy based on the Democracy Index 2016, a report by The Economist Intelligence Unit. This analysis will help to answer the following questions by John Elkington:

1. Can we vote our way to a sustainable future for a world of 9 billion people?
2. Can the time-frames of democratically elected governments deliver sustainable development?

The author, for the sake of this paper, is making an assumption that the answers to Elkington's other questions ("are new forms of leadership going to be necessary?" and "what kinds of systems of governance and decision-making best place countries and people to move towards more sustainable forms of development?") are as follows:

1. No new forms of leadership are going to be necessary;
2. Democratic system of governance based on pluralistic forms of ensuring the social and democratic balance is sufficient to best place countries and people to move towards more sustainable forms of development.

The Economist Intelligence Unit's Democracy Index provides a snapshot of the state of democracy worldwide for 165 independent states and two territories. This covers almost the entire population of the world and the vast majority of the world's states (microstates are excluded). The Democracy Index is based on five categories: electoral process and pluralism; civil liberties; the functioning of government; political participation; and political culture. Based on their scores on a range of

Table 1 Democracy Index 2016, by regime type (The Economist Intelligence Unit 2016)

	No. of countries	% of countries	% of world population
Full democracies	19	11.4	4.5
Flawed democracies	57	34.1	44.8
Hybrid regimes	40	24.0	18.0
Authoritarian regimes	51	30.5	32.7

Note "World" population refers to the total population of the 167 countries covered by the Index. Since this excludes only micro states, this is nearly equal to the entire estimated world population

indicators within these categories, each country is then itself classified as one of four types of regime: "full democracy"; "flawed democracy"; "hybrid regime"; and "authoritarian regime". The Democracy Index (2016). Considering that you can implement pluralistic forms of ensuring the social and democratic balance only in full democracies and maybe in flawed democracies, the first thing to do is to analyze how many of the world's regimes are in one of those groups. That information is presented in Table 1.

As we can see, 45.5% of countries are qualified either as full or flawed democracies. That translates to 49.3% of global population. If we take a closer look as to which of the countries are considered to be full or flawed democracies, we will be able to determine what part of global economy they control. That information can be found in Fig. 1 that shows the regimes by region.

When we compare the data from graphic 1 with the data from Table 2 presented below we will have a more transparent image of the economic power of democratic regimes around the world.

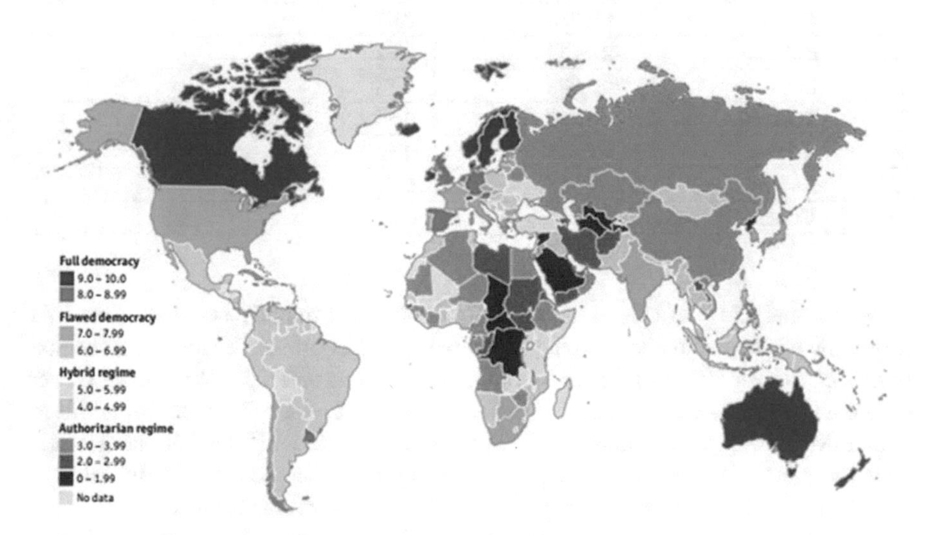

Fig. 1 Democracy Index 2016, democracy around the regions (The Economist Intelligence Unit 2016)

Table 2 Participation in the global economy in 2015 by country (World Bank 2015)

No.	Country	GDP in mln USD	Participation in the global economy (%)
1	USA	17,946,996	24.44
2	China	10,866,444	14.80
3	Japan	4,123,258	5.61
4	Germany	3,355,772	4.57
5	Great Britain	2,848,755	3.88
6	France	2,421,682	3.30
7	India	2,073,543	2.82
8	Italy	1,814,763	2.47
9	Brazil	1,774,725	2.42
10	Canada	1,550,537	2.11
11	South Korea	1,377,873	1.88
12	Australia	1,339,539	1.82
13	Russia	1,326,015	1.81
14	Spain	1,199,057	1.63
15	Mexico	1,144,331	1.56
16	Indonesia	861,934	1.17
17	Netherlands	752,547	1.02
18	Turkey	718,221	0.98
19	Switzerland	664,738	0.91
20	Saudi Arabia	646,002	0.88
21	Argentina	548,055	0.75
22	Sweden	492,618	0.67
23	Nigeria	481,066	0.66
24	Poland	474,783	0.65
25	Belgium	454,039	0.62
26	Iran	425,326	0.58
27	Thailand	395,282	0.54
28	Norway	388,315	0.53
29	Austria	374,056	0.51
30	Venezuela	371,337	0.51

When analyzing data from the presented sources it quickly becomes clear that just in the top 10 democracies in the world we have a combined economic force that translates to over 50% of global economy. If we add to that the fact of rising power of Indian economy and we assume that the President of the People's Republic of China, Xi Jinping, was serious when he stated that China will be a fully democratic state by 2050, then we are looking at a majority of global economy being able to adopt pluralistic forms of ensuring the social and democratic balance.

With that being said and with the above analysis in mind, the questions of John Elkington can be answered as follows:

1. Yes, we can vote our way to sustainable future. Of course, not all of the 9 billion people will be able to do so but the majority that controls most of the crucial resources will be able to do so and to lead by an example.
2. The time-frames of democratically elected governments can deliver sustainable development if they adopt the pluralistic forms of ensuring the social and democratic balance.

The author realizes that those positive answers are only a theoretical concept and he is not so naive as to claim that those concepts are easily implemented in real life scenarios. Even still it is encouraging and refreshing to realize that there are different ways of building a sustainable future in which the society is able to decide on their own behalf, build the world of safe, constant growth and socio-democratic balance.

References

Bartolini S, Mair P (1990) Identity, competition and electoral availability. The stabilization of European electorates 1885–1985. ECPR Press, Colchester

Dahl R (1956) A preface to democratic theory. University of Chicago Press, Chicago

Dahl R (1972) Polyarchy: participation and opposition. Yale University Press, New Haven

Joseph J (2001) Sustainable development and democracy in the megacities. Dev Pract J 11(01)2–3: 102–121

Laneve G (2017) Environmental education for sustainable development and active citizenship participation. Italian experience. Paper presented at 7th annual conference on political science sociology and international relations, GSTF, Singapore, 24–25 Sep 2017

Tłokiński W, Wiśniewski B (2017) Committee for the defence of democracy (KOD): pluralistic form of ensuring the democratic balance in Poland. Paper presented at 7th annual conference on political science sociology and international relations, GSTF, Singapore, 24–25 Sep 2017

The Economist Intelligence Unit: The Democracy Index (2016) Revenge of the "deplorables". http://www.eiu.com/topic/democracy-index. Accessed 25 Sep 2017

The Environment Foundation: http://www.environmentfoundation.net/. Accessed 26 Sep 2017

World Bank (2015) Open Data. https://data.worldbank.org/. Accessed 27 Sep 2017

Cultural Landscape of Żuławy Wiślane in the Light of the Idea of Sustainable Development

Klaudia Nowicka

Keywords Cultural landscape · Sustainable development · Żuławy Wiślane

1 Introduction

During the 20th century, unconditional growth and development in conjunction with environmental degradation stimulated a global commitment to landscape preservation. Protecting both the natural and the cultural environment has become a global issue, and it is reflected in many international treaties and declarations, such as UNESCO or RAMSAR. In this article the idea of sustainable development is to be used in the context of the cultural landscape of a region of Żuławy Wiślane, which is considered to be unique as being a result of several stages of settlement and influenced by people representing different backgrounds and cultures living and working there over time. When, finally, culture was recognised as the fourth pillar of sustainable development, a wide range of approaches, methods and tools of environmental analysis have become available for analysing cultural landscapes, as there are many parallels between natural and cultural sustainable development. This article is an attempt to perform a mainly descriptive analysis of the cultural landscape of Żuławy Wiślane according to the categories of the STAMEX system proposed by Solon (2004). In order to do so, the author analysed official materials, maps and strategic documents concerning the analysed region. She also carried out fieldwork in July and August 2017 during which several study visits and interviews with representatives of organizations and activists involved in the protection of the cultural landscape of Żuławy Wiślane were made.

K. Nowicka (✉)
University of Gdansk, Gdansk, Poland
e-mail: klaudia.nowicka@ug.edu.pl

T. Zielinski et al. (eds.), *Interdisciplinary Approaches for Sustainable Development Goals*, GeoPlanet: Earth and Planetary Sciences,
https://doi.org/10.1007/978-3-319-71788-3_6

2 The Idea of a Cultural Landscape

The term "cultural landscape" emerged in the 19th century and although it was used by German, French and American geographers, it was variously understood. German researchers paid more attention to its evolution and visual aspects, while in French geography the main focus was on determining relations between human activities and the environment that is visible in the landscape. In the USA, geographers considered the cultural landscape as one continuous whole—they did not divide it into natural and anthropogenic landscapes. It was a region with typical relations between natural and cultural elements which are the result of human activity in the environment (Kowalczyk 2008). In Polish literature of the 20th century, cultural landscape was not identified with a region (Dobrowolska 1948), and many other terms emerged, for instance, the "cultural environment" (Czyżewska and Kostarczyk 1989; Medowski 1989) or "cultural space" (Kołodziejski 1989). In this paper the American approach prevails, considering the whole region of Żuławy Wiślane as the cultural landscape.

In the 1970s Lewis (1979) formulated several theses on the cultural landscape and its interpretation. According to him, a particular cultural landscape is an imprint of people and groups on the land—an evidence of their culture. In this context, cultural landscapes can be read as books providing knowledge on people who co-create the landscape they exist in. Moreover, P. Lewis claimed that all elements of the landscape should be equally treated and investigated in the context of historical conditions where a particular landscape has emerged. He claimed that the "[…] human landscape is our unwitting autobiography, reflecting our tastes, our values, our aspirations, and even our fears, in tangible, visible form" (Lewis 1979, 12). Therefore, it can be perceived as a kind of monument commemorating a given group of people. According to Homans (1942, 1969), a cultural landscape is a record of the history of a given social group living in a particular environment. This history is written in the form and structure of the landscape. A French geographer, Paul Vidal de la Blache, perceived culture as a source of ideas, values, customs and beliefs, which are all basis for a destructive or creative approach to the natural environment surrounding human beings. According to de la Blache's concept, the external environment (fr. *milieu externe*) determines boundaries for human actions while the system of values and ideas (fr. *milieu interne*) sets dynamics and directions of those actions. The cultural landscape (fr. *paisage humanise*) emerges at the melting point of *milieu* and the so-called style of living (fr. *genre de vie*) and it reflects the way particular groups of people interpret and use their environment. P. Vidal de la Blache also claimed that the landscape is a result of history and culture which both determine ways of life and relations between humans and nature (Buttimer 1978; Rembowska 2002; Wilczyński 2005). The impact that societies exert on the landscape surrounding them reflects in the names given to particular places, visual symbolism and structures. Thus, the norms of behaviour adopted by a given group of people are visible in the landscape it creates (Butzer 1978). However, as Mitchell et al. (2014, 117) states, "[…] the complex array of cultural and natural values represented by tangible and intangible heritage are not readily understood by the outside visitor or manager. Understanding requires honoring the world-views and core values of the communities that are (or were) their stewards over different periods, listening to diverse voices and perspectives, and respecting the different knowledge systems and practices embedded in these places, which have much to teach us about resilience."

In the situation when alien cultures meet, a process of cultural landscape rejuvenation begins, or the new cultural landscape emerges on the remains of the former one. The region of Żuławy Wiślane is such a multilayered cultural landscape with visible elements created over time by different groups of people. There are also initiatives aimed at understanding tangible and intangible elements of the cultural landscape of Żuławy Wiślane which has emerged as the result of a long and complex relationship between nature and different communities living and working there. The most unusual group of people, which exerted tremendous influence on the cultural landscape of Żuławy Wiślane, lived there from the second half of XVI century to 1945. Those people were Mennonites and their cultural heritage is to be analysed and described in the greatest detail.

3 Sustainable Development of Cultural Landscapes

From the Industrial Revolution to the second half of the 20th-century culture (no matter how it was defined) was not associated with development. It was rather seen as something hampering development or even making it impossible (Streeten 2011). During the last decades a visible change in perceiving culture has been observed. It started to be recognized as a factor playing an important role in the development processes, especially in the context of the wider perspectives—socio-economic development and, primarily, sustainable development in which economic growth is vital, yet not the only one aspect of development (Palmer and Purchla 2010). Lately, the three basic dimensions of sustainable development—ecological, economic and social—have been supplemented by the fourth one: cultural dimension, which is now perceived as an important pillar or even a vital driving force of sustainable development. It has become clear that the processes of initiating and implementing changes as well as the process of governance may be culturally conditioned (Our Creative Diversity 1995; World Bank 1999; Cernea 2001; Hawkes 2001; Greffe et al. 2005; Pascual 2009; Janikowski and Krzysztofek 2009; Myga-Piatek 2010; Throsby 2010; From Green Economies … 2011; Streeten 2011; Murzyn-Kupisz 2013).

Throsby (1995, 2003, 2010) claims that culturally sustainable development may be described using and preserving the same principles that have been established in regard to natural resources, especially those created in the field of environmental economics and ecological economics. At the same time, Throsby refers to the concept of capital, not only in its narrow meaning, but also to other forms of capital like social capital, natural capital and cultural capital. When considering the issues related to sustainable development in the context of cultural heritage perceived as the above-mentioned cultural capital, it becomes obvious that the same principles and mechanisms of sustainable development can be applied as those regarding natural values (natural capital) (Navrud and Ready 2002). Both natural and cultural resources are the heritage of the past, no matter if they have been made by nature or people. However, they are undergoing constant changes and each generation decides what is worth preserving and conserving. That is why the basic principles of sustainable development which are intergenerational and intragenerational equity as well as the precautionary principle when managing non-renewable resources (Harding and Fisher 1995) may apply to elements of cultural heritage as well. "Recognition of cultural landscapes has also influenced the theory and practice of historic preservation and its relationship to nature conservation. Cultural landscape conservation has shaped a concept of heritage that has become increasingly dynamic and inclusive. Importantly, conservation of many cultural landscapes is reliant on local and indigenous leadership and governance as well as traditional knowledge systems and institutions, and it is integrated with other policies and programs" (Mitchell et al. 2014, 117). This approach is also visible in many international studies and documents, for instance, in the Report of the World Commission for Culture and Development entitled '*Our Creative Diversity*' (Our creative diversity 1995). Three years later, in 1997, a report '*In from the margins—A contribution to the debate on Culture and Development in Europe*' was published in which the development potential of cultural heritage was recognised (Murzyn-Kupisz 2013). Then, in 1998, UNESCO deemed cultural policy as a key constituent of the development strategy in the report '*The Power of Culture*' (UNESCO 1998). The coexistence of natural and cultural resources and their equal importance have been emphasised in the preamble to the European Landscape Convention established in 2000 by the Council of Europe (European Landscape Convention 2017): 'The landscape has an important public interest role in the cultural, ecological, environmental and social fields, and constitutes a resource favourable to economic activity and […] the landscape contributes to the formation of local cultures, and it is a basic component of the European natural and cultural heritage, contributing to human well-being and consolidation of the European identity.' In the UNESCO 2005 *Convention on the Protection and Promotion of the Diversity of Cultural Expressions*, there are direct references to the concept of sustainable development. The Preamble to the Conventions emphasises 'the need for incorporate culture as a strategic element in national and international development policies, as well as in international development cooperation […]' and the fact that 'cultural activities, goods and services have both an economic and cultural nature.' Among the guiding principles, there are the 'Principle of the complementarity of economic and cultural aspects of development'; the Principle of sustainable development', saying that 'cultural diversity is a rich asset for individuals and societies. The protection, promotion and maintenance of cultural diversity are an essential requirement for sustainable development for the benefit of present and future generations' (Convention on the Protection … 2017). Finally, from the factor hampering development to the key constituent of the development strategy, culture has come a long way to become the fourth pillar of sustainable development.

4 Location and Divisions of Żuławy Wiślane

Żuławy Wiślane is a region in northern Poland situated in the Vistula Delta. It is a plain stretching longitudinally between the Vistula Spit and the point where the rivers Nogat and Vistula meet (Iława Lake District and Kwidzyń Valley); latitudinally, it stretches from the Kashubian Lake District and the Starograd Lakeland to the Elbląg Upland and the Warmia Plain. According to the physico-geographical regionalisation by Kondracki (2000), Żuławy Wiślane is a mesoregion, being a part of the macroregion named Pobrzeże Gdańskie. Żuławy Wiślane covers the area of approximately 175 thousand hectares, including 45 thousand hectares of depression areas. The lowest point is in Raczki Elbląskie where the depression reaches 1.8 m below the sea level.

Two main rivers flowing through the region are the above-mentioned Vistula and Nogat which internally divide Żuławy Wiślane into Żuławy Gdańskie, Wielkie Żuławy Malborskie, Małe Żuławy Malborskie and Żuławy Elbląskie (Fig. 1).

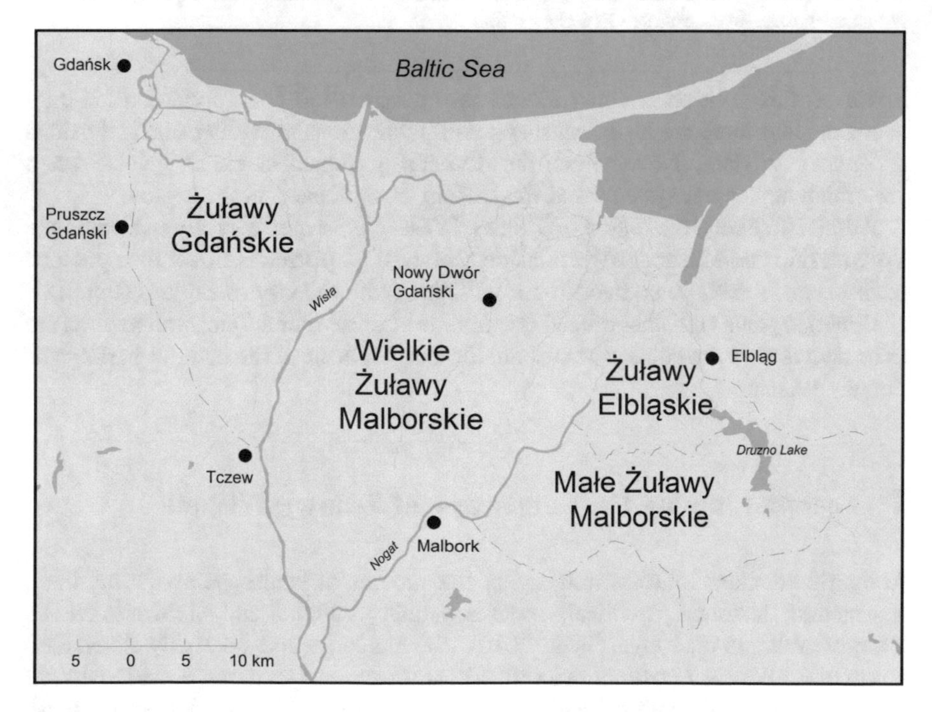

Fig. 1 Location and conventional division of Żuławy Wiślane. *Source* Own elaboration on the basis of Żuławy Info (2017)

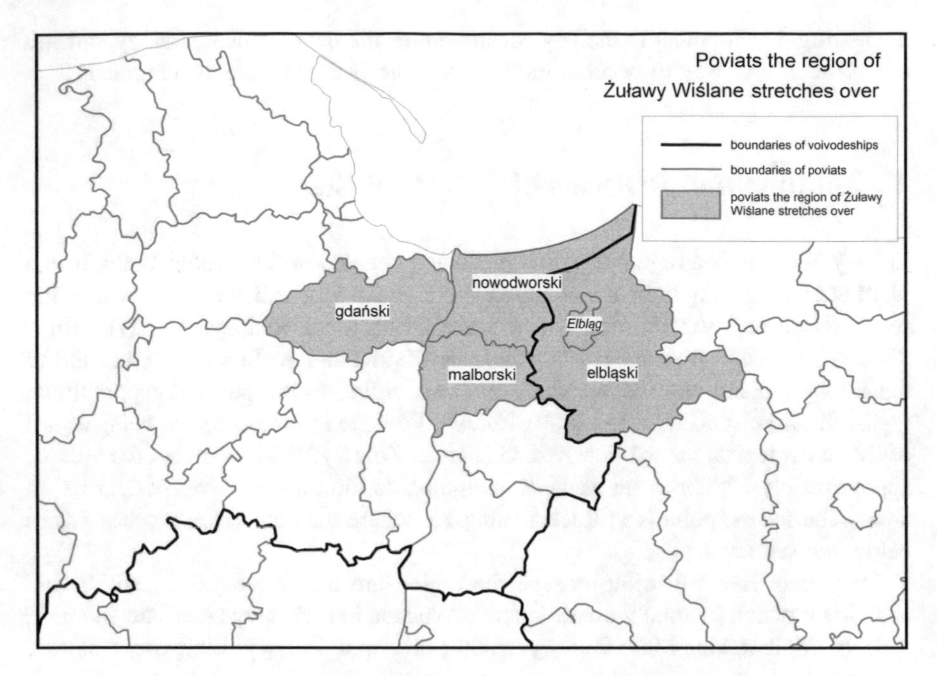

Fig. 2 Poviats of Pomorskie and Warmińsko-Mazurskie Voivodhips the region of Żuławy Wiślane stretches over. *Source* Own elaboration

However, this division is conventional as the region has been divided differently over time and there are historical maps and publications providing other divisions of Żuławy Wiślane. Nonetheless, the division presented herein is helpful when describing settlement systems and the history of settlement in the region.

Administratively, the region of Żuławy Wiślane is located partially in Pomorskie Voivodeship and Mazursko-Warmińskie Voivodship. It stretches over four poviats: gdański, malborski, nowodworski and elbląski (with the city of Elbląg) (Fig. 2).

Strategic plans and other official documents of those four administrative units are to be analysed in the context of sustainable development of the cultural heritage of Żuławy Wiślane.

5 Factors Shaping the Landscape of Żuławy Wiślane

Among fundamental factors influencing the process of landscape evolution, there are natural, historical, political, socio-economic, cultural and civilization ones (Dobrowolska 1948; Myga-Piątek 2010). Such a large and internally diversified group of factors may pose serious difficulties for general analyses. That is why, in most cases, they are analysed and interpreted on a regional scale. Moreover, those factors are subject to constant changes; consequently, they cannot be fully explained and analysed solely on the grounds of environmental determinism or environmental nihilism. The idea of possibilism, introduced by the already-mentioned Paul Vidal de la Blache (Buttimer 1978), has provided a new way of holistic and complementary understanding of the factors themselves and the role they play in the life of societies and processes of landscape evolution. Thus, the idea of Vidal de la Blache was a foundation for the broadly-understood sustainable development which is, according to Pawłowski (2009), the harmonic development of societies based on respect for natural laws and cultural achievements.

5.1 Natural Factors

Natural factors include geological structure, relief, water regime, climate, soil, fauna and flora. According to Friedrich Ratzel and the concept of geographical determinism, those factors determine the conditions of settlement and civilization development. They may ease or hamper economic development or even make it impossible (Eberhardt 2015; Myga-Piątek 2010). Taking the main subject of this article under consideration, the natural factors shaping the landscape of Żuławy Wiślane are not to be elaborated in great detail although they have been affecting human efforts to transform the delta into an arable and liveable land.

Żuławy Wiślane is the youngest part of the land in Poland which has been formed from the deposition of sediment carried by the Vistula River. Not later than 16 thousand years ago, the Scandinavian glacier retreated from this area. The emergence of the Baltic Sea triggered a process of building the delta. Approximately 10 thousand years ago the Baltic Sea level was lower by 100 m and it was rising for about 4 to 5 thousand years when it reached more or less the level it is at present; the slope of the Vistula in its estuarial zone decreased to 0.05% and the process of delta forming began. As a result of waves, currents, winds and river activities, an underwater sandbank was created in the shallow waters of the Gdansk Bay, which then emerged from the water, forming the Vistula Spit and the Vistula Lagoon. For many years the lagoon covered a large part of the area of today's delta. From this time on, the delta was being created as an internal delta. Sand, sludge and clay—materials carried by the Vistula and Nogat—were settling at the bottom of the Vistula Lagoon and making it increasing shallower. Over time, the process of delta formation slowed down and then it was practically stopped by human activity which started in the 13th century. By the end of the 19th century, the flow of the Vistula and Nogat was finally regulated and the process of delta building was over (Strupińska 2001; Szafran 1981; Lipińska 2011).

Apart from the two main rivers—the Vistula and Nogat—surface waters of Zulawy Wislane include other rivers: Szkarpawa, Elbląg, Motława, Radunia, Tuga, Święta, Tina and other smaller ones. All those rivers have minimum slopes and their level almost equals the sea level. Most of the time, the water flows smoothly, but on windy days the areas at the mouth of those rivers are prone to flooding. Sometimes the backflow can be observed. The only lake in Żuławy Wiślane is the

Druzno Lake (Fig. 1). This lake is the remnant of the large lagoon. Nowadays, this water body covers the area of approximately 1790 ha and its depth is only about several meters. The lake is largely overgrown and it is a habitat for many species of birds (Szukalski 1975). Apart from the already-mentioned natural water bodies, there are many anthropogenic ones which are to be mentioned later when describing the civilization and cultural factors.

The region of Żuławy Wiślane is characterised by high air and soil humidity resulting from shallow ground water placement, a dense network of drainage ditches and proximity to the sea. All those elements affect the air temperature too which is cooler than inland. Żuławy Wiślane is surrounded by moraine hills and the eastern and western uplands are natural barriers for winds. That is why the region has a lower precipitation than the surrounding areas (Szukalski 1975).

In the past, Żuławy Wiślane was covered by lush riparian forest. Nowadays, there are hardly any woodlands in the region. The oldest cluster of trees is a narrow strip of Mątawski Forest at the fork of the Vistula and Nogat Rivers (Fig. 1). In the remaining area there are only single old oaks left, yet willows are the trees perceived as a landmark of Żuławy. There are four sanctuaries in the region: Mątawski Forest, Druzno Lake, Nogat Estuary, Nogat Estuary Protected Landscape Area, and most of them are habitats of rare bird species (Fig. 3).

The Druzno Lake is one of the thirteen sites in Poland protected under the Ramsar Convention on Wetlands of International Importance, especially as Waterfowl Habitat. The most widespread aquatic vegetation is represented through floating communities of different associations of water lilies. The site is important

Fig. 3 The Druzno Lake. *Fot* Klaudia Nowicka

for birds migrating along the Baltic coastline and provides refuge for more than 150 bird species during the summer (Ramsar Sites Information … 2017).

Żuławy Wiślane is a region where the natural factors dominated over and determined human activities for many years. Numerous floods, undermining human labour and technical inventions, have always served as a reminder of how powerful nature can be.

5.2 *Historical and Political Factors*

Historical and political factors are all events resulting from border changes, wars, administrative divisions, systems of power, political regimes and legal systems. They all have left their mark on the landscape and its spatial organisation (Myga-Piątek 2010).

The first traces of settlement in Żuławy Wiślane dates back to the 2nd century BC and they were found near Kaczynos and Janowo villages. Around the year 1000 BC there were three important settlements in the region: Truso (at the mouth of the Nogat river), Gyddanyzc (Gdansk) and Neuteichsdorf (Nowy Staw). However, the most intense settlement process started in 1308–1309 when the Grand Master of the Teutonic Order moved from Venice to Malbork. The Kulm (Chełmno) Law was a legal basis for the location of new villages at that time. During this stage of the settlement process, which lasted until the 15th century, new settlement units emerged mainly in Żuławy Malborskie in the southern part of Żuławy Gdańskie (Fig. 1). The Kulm Law was in force until the Partitions. The next stage started when the Order lost in the Thirteen Years' War (1454–1466) and Żuławy Wiślane became Polish. During this period, in the 16th century, settlers from the Netherlands arrived and settled in the areas which had not been meliorated yet. They were representatives of the Anabaptist movement which started in Switzerland during the time of the Reformation. The movement rapidly gained popularity in Europe, unlike their ideas which were opposed to the traditional relationship between church and state. As a consequence the new believers were persecuted, which led to a mass migration of the Anabaptists. In Poland Mennonite congregations found business opportunities, land and religious tolerance. In Żuławy Wiślane they were settling near Gdańsk (Żuławy Gdańskie) and in Żuławy Malborskie and Żuławy Elbląskie. The Partitions of Poland (1772–1795) started the next stage of the region's history as it became a part of East Prussia. The Mennonites, whose religion prohibits use or possession of weapons, came into conflict with the new ruler Frederick Wilhelm II. Some of them left Żuławy and moved to Ukraine. However, this was the period when Nowy Dwór Gdański became an important trade centre. During the interwar period, by virtue of the Treaty of Versailles, the Free City of Danzig was established and it covered the areas of Żuławy Gdańskie and Żuławy Malborskie while the region of Żuławy Elbląskie was still part of East Prussia. At the end of the Second World War, in March 1945, the withdrawing German army undermined the dikes and dams. The whole region flooded and turned into wetlands and marshes—

it was totally cut off from Gdansk and ruined. After three years of hard work, in 1948, Żuławy Wiślane was finally dry and again new settlers arrived—repatriates from the Soviet Union and people from central Poland. They settled in randomly chosen villages and settlements. The post-war period introduced a new political system—socialism with the economy based on large-scale agriculture. Finally, the 1980s crisis undermined the efforts of people who had been trying to save Żuławy. After the fall of socialism, the region was given a second chance, and people who had been born there started to identify themselves with Żuławy Wiślane and its rich heritage (Szukalski 1975; Opit 1998; Lipińska 2011; Mężyński 1960–1961; European Mennonite Network 2017). The region, although still associated mainly with good soils and agriculture, had finally regained its multilayered cultural identity.

5.3 Cultural Factors

Cultural factors are a manifestation of the gradual cultural evolution of societies. They are ways of building homes and other objects, architectural styles, scientific discoveries, technical inventions, a growing sense of territorial identity, spiritual culture: language (local dialects), customs, traditions, rites, beliefs, religion (Myga-Piątek 2010).

Each of the numerous stages that have already been distinguished in the region's history has left its mark, which is visible in elements of tangible and intangible cultural heritage of Żuławy Wiślane. The most unusual—for Poland—elements of cultural heritage are related to the period when the Mennonites used to live in Żuławy: from the second half of the 16th century to 1945 when those who had not moved to Ukraine earlier were killed by the Red Army soldiers.

The spiritual culture of this group is mainly related to their religion. The Mennonites are Christian groups belonging to the church communities of Anabaptist denominations named after and influenced by the teachings and traditions of Menno Simons of Friesland, who is the greatest character in the history of the Mennonite Church. The Word of God written in the Bible was the foundation of the Mennonite religion, and the Bible is the only authority on belief and human life. Mennonites believe the Bible is divinely inspired and that Jesus Christ died on the cross to save humanity from its sins. Mennonites believe "organized religion" is important in helping individuals understand their purpose and in influencing society. Church members are active in serving the community, and a large number participate in missionary work. The church places a strong emphasis on a Christ-centred life and service to others as well as a repudiation of infant baptism and capital punishment. Mennonites were prohibited from holding public office and taking any oaths. The church has long held a belief in pacifism. Members act this out as conscientious objectors during war but also as negotiators in resolving conflicts between warring factions. Most Mennonites were iconoclasts and did not allow anybody to photograph them. As they did not allow strangers to take part in their ceremonies, there is virtually no iconographic material related to this religious group (European Mennonite Network 2017). Numerous objects documenting spiritual the culture of the Mennonites living in Żuławy Wiślane can be found in the museum Żuławyski Park Historyczny in Nowy Dwor Gdański.

The tangible elements of the cultural heritage in the Żuławy Wiślane landscape are the result of the multi-stage settlement process with visible medieval, gothic, Teutonic, Mennonite and other influences. The cultural heritage comprises numerous sacred objects (churches, cemeteries), windmills, famous arcaded houses, hydrotechnical objects and the Castle of the Teutonic Order in Malbork. Among the most interesting sights there are numerous Mennonite cemeteries [e.g. Stogi, Stawiec, Szaleniec, Markusy (Fig. 4)], churches in Wróblewo, Nowy Staw, Lichnowy and Jezioro, the unique arcaded houses in Miłocin and Trutnowy (Fig. 5), the "Gdanska Glowa" lock in Drewnica and windmills in Drewnica and Palczewo.

An interesting object of cultural heritage of Żuławy (or more strictly Nowy Dwór Gdański) is Stobbes Machandel. It is juniper vodka produced from 1776 to 1945 by a Mennonite family Stobbe. It became very popular in Gdansk, especially during the Free City period. Drinking Stobbes Machandel involved a special ritual. First, to an empty glass a dried prune was put on a toothpick, then vodka was poured on top. Then one had to eat a plum, keeping the plum stone in the cheek. In the left hand one had to hold a toothpick. Then in one gulp vodka was drunk (without swallowing the plum stone). Then one would spat the pit into an empty glass. Later, referring to the Hanseatic habit of breaking masts, the toothpick was broken in half and left in the glass.

Fig. 4 Cemetery in Markusy. *Photo* Klaudia Nowicka

Fig. 5 The largest arcaded house in Żuławy Wiślane (Trutnowy). *Source* Mierzeja i Żuławy (2017)

5.4 Socio-Economic Factors

Socio-economic factors include settlement systems, forms of land and material goods ownership, and occupational and social structures (Myga-Piątek 2010).

Żuławy Wiślane is a region rich in various settlement forms and village layouts. All main types of settlement patterns can be found in the landscape. There are all kinds of nucleated villages as well as dispersed settlements. A characteristic feature of the settlement landscape of the region is the so-called "terp"—a natural or human-made hill on which houses and other buildings were situated. *Terps* are the visible proof of how people fought with water and how the natural features determined their everyday lives. Numerous floods forced people to adapt the architecture of their houses to the environmental conditions—the famous arcades were built to protect the crops. Another interesting and distinctive feature of the region is the field layout which owes its uniqueness to melioration ditches surrounding each field and making it easy to distinguish between single farmland areas. As Żuławy Wiślane is a plain with almost no forests, houses and other buildings dominate the landscape, creating an extraordinary picture. The uniqueness of the Żuławy Wiślane landscape made Maria Kiełczewska-Zalewska (1956), who studied the settlement patterns in the whole Pomerelia, to exclude this region from her study.

As for occupational and social structure, from the very beginning, thanks to fertile soils and numerous water bodies, Żuławy Wiślane has been occupied by farmers and fishermen. Among the already-mentioned consecutive waves of migrants and repatriates, the most interesting group was Mennonites who have affected the cultural landscape the most. Since they were very hardworking and had remarkable meliorating skills and knowledge, Mennonites became very rich and influential in the relatively short term. They built large houses and mills that are still present in the landscape of Żuławy.

5.5 Civilization Factors

Civilization factors described by the intellectual and biological potentials of societies, access to technical achievements and material goods. Those factors have increased people's feeling of protection and gradually eased relations between the environment and human beings, finally leading to environmental nihilism (Myga-Piątek 2010).

In the case of Żuławy Wiślane, the achievements of the implemented hydraulic engineering and the water engineering structures visible in the landscape and protecting people from the element water are civilization factors that had tremendous influence on the present shape of the cultural landscape. There are numerous water watchpoints, pumping stations, locks, weirs, culverts, storm gates and drawbridges, e.g. in Kiezmark, Różany (Fig. 6), Osłonka and Rybina.

Fig. 6 Historic pumping station in Różany. *Photo* Klaudia Nowicka

The variety of different structures and objects makes the landscape of the region unique and tells the history of people fighting nature.

All the factors, from natural to civilization ones, and their multilayered relations have created a unique—for both Poland and Eastern Europe—cultural landscape that is extraordinary in natural and anthropogenic dimensions. In its apparent monotony, triggered by a striped layout of the landscape (Lipińska 2011), one can find a wide range of tangible and intangible objects of cultural heritage, created by an exceptional mixture of people who have lived there, in the shadow of the element water.

6 The Level of Landscape Sustainability Measured Using the STAMEX System

Numerous sets of indicators have been established for the purpose of landscape assessment and sustainable development measurements. Among the widely used ones are systems recommended by the OECD and UN Commission on Sustainable Development. They are mainly based on the concept involving three separate categories of phenomena occurring at the junction of human activity and natural environment: pressure, state and response (Solon 2004). However, as Solon (2004) claims, it is very problematic to distinguish between the pressure and the actual state of the environment. Therefore, he proposes the implementation of a new system: state-model-execution (STAMEX) in which establishing a model of sustainable landscape is the key yet challenging issue. Although Solon (2004) refers to the environmental dimension of sustainable development, it is possible, as already mentioned, to adopt the same assessment methods as there are many parallels between naturally and culturally sustainable development (Thorosby 1995, 2003, 2010). In case of cultural heritage, establishing the model of a culturally sustainable landscape is not as complicated as there are official documents describing tangible and intangible elements of the cultural landscape being under protection and worth preserving for future generations. This article is an attempt to perform a mainly descriptive analysis of the cultural landscape of Żuławy Wiślane according to the categories of the STAMEX system proposed by Solon (2004).

6.1 Model

The model of the sustainable cultural landscape of Żuławy Wiślane includes numerous elements of tangible and intangible elements of cultural heritage. The model shall include objects of historical value, original settlement patterns and field layouts as well as elements of spiritual culture representing all settlement stages and peoples which have already been described when briefing the historical and political factors shaping the landscape of the analysed region. They are as follows: (1) the pre-Teutonic stage, (2) the Teutonic stage, (3) the Mennonite stage, (4) the Prussian stage and (5) the contemporary stage. Lipińska (2011) in her study on preserving and shaping the cultural landscape of Żuławy Wiślane determined 22 model historical-landscape units representing the settlement stages 1–4. However, as it was stated above, the model sustainable cultural landscape shall comprise elements representing all of them, including modern objects (e.g. wind farms affecting the landscape of Żuławy Wiślane) and traditions while respecting the principle of intergenerational and intragenerational equity as well as the precautionary principle.

6.2 State

In order to assess the current state of the cultural landscape of Żuławy Wiślane, it is necessary to divide this broad concept into two dimensions: tangible and intangible ones, as in the case of the analysed region the condition of those dimensions substantially differs and needs to be described and assessed separately. In a great simplification, the first dimension comprises tangible elements of culture and the second one tradition and other elements of spiritual culture. Although, in most cases, those two dimensions are tightly connected, Żuławy Wiślane is extraordinary in this context, too.

In her study of 2011, B. Lipińska distinguished 26 regions of different sizes and ranges representing the already-mentioned model historical-landscape units. The presence of so many differentiated units confirms that the monotony of the cultural landscape of Żuławy Wiślane is indeed illusive and that the region is strongly internally differentiated. The distinguished regions reflect the influence of various natural, historical and political, socio-economic, cultural and civilization factors. For instance, the region of Vistula Spit (Region I) and the areas along the Nogat and Vistula rivers (Regions XVII and XVIII) comprise unchanged elements of the natural landscape; the regions: Sobieszewo-Sztutowo (Region II), Szkarpawa (Region IV), areas along the Motława River (Region VII), Gozdawa and its vicinity (Region X), a group of the 17th century villages "Pięć Grobli" including Janowo, Józefowo, Adamowo, Kazimierzowo and Władysławowo (Region XII), Jegłownik and its vicinity (Region XIV), Lake Druzno and its surrounding areas (Regions XV and XVI), Kącik village (Region XX), Stogi village (Region XXIII), Mechnica and Mojkowo villages (Region XXVI). These all comprise elements of Mennonite culture. The regions representing the medieval culture and period when the Teutonic Order had power over Żuławy Wiślane are the region of villages located in Żuławy Malborskie Wielkie and Żuławy Gdańskie (Region IX), Grabiny-Zameczek (Region XXI), Skowronki and Kałdowo (the suburbs of Malbork) (Region XXIV), Jegłownik (Region XXV). Some visible results of the civilization factors' influence in the shape of hydrotechnical objects and structures are situated in the regions where there used to be or still are windmills and other drainage facilities (Region II:

Sobieszewo-Sztutowo, Region IV: Szkarpawa, Region VI: Olszynka, Region XI: areas surrounding the Nogat River and its tributaries, Region XIV: Jegłownik and its vicinity, Regions XV and XVI: Lake Druzno and its surrounding areas, Region XX: Żuławki, Drewnica, Przemysław, Izbiska.

Lipińska (2011) assessed the percentage/level of preservation of the historical landscape of villages, towns and cities situated in Żuławy Wiślane. The average percentage of all 254 settlement units located in the four poviats covering the analysed region is 62%, with the highest level of preservation of the historical landscape (90–100%) in Stobna, Markusy, Kępniewo and Chorążówka (Fig. 7). The lowest level of preservation of the historical landscape (0–10%) was recorded in Jankowo Żuławskie, Stawidła, Stegna, Wybicko, Stare Babki, Włodarka, Zadworze, Gronowo Elbląskie, Klimonty, Kałdowo, Dziewięć Włók and Widowo. Among them are settlement units which have been completely destroyed (e.g. Widowo), modernised and transformed [e.g. as a result of constructing railway like in the case of Gronowo Elbląskie or as a result of implementing large-scale farming solutions—the state-run farms called PGRs (Włodarka, Zadworze, Stare Babki, Jankowo Żuławskie, Wybicko)], turned into modern suburbs (Kałdowo) or changed under pressure of developing a tourist function (e.g. Stegna). Yet, in 177 out of 254 towns and villages, there are still objects of historical and cultural value although many of them are in poor condition.

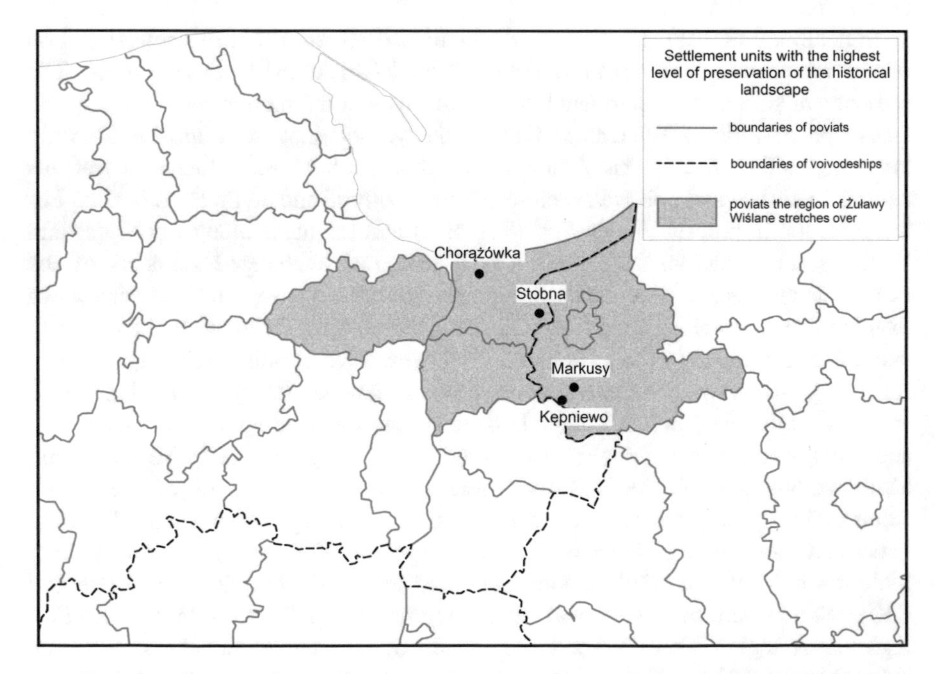

Fig. 7 Settlement units with the highest level of preservation of the historical landscape. *Source* Own elaboration

The historical tradition of Żuławy Wiślane, the region whose culture has been shaped by many groups of people with different backgrounds and cultures, no longer exists as the most important factor and its driving force is missing: a human-resident-creator who has been deeply attached to the land for many successive generations. Nowadays, there are practically no descendants of societies living and working in Żuławy Wiślane in the past—they were all displaced or killed in 1945 (Lipińska 2011). This lack of integrated community turned the cultural tradition into something formal, based only on external manifestations with no lively and creative content. Many objects documenting the spiritual culture of the former societies living in Żuławy Wiślane can be found in museums in Nowy Dwór Gdański, Elbląg and Malbork. However, there are some signs that people currently living in the region, being under the influencc of Żuławy Wiślane and its unique character, are creating their own cultural heritage. This new tradition is also influenced by the natural factors, especially the element water, so it resembles the historical one in many dimensions. It seems that the environmental nihilism is not the approach which shall be associated with Żuławy.

6.3 Execution

All efforts which have been made to make the cultural landscape of Żuławy Wiślane sustainable can be divided into two basic categories: top-down and bottom-up initiatives.

The top-down initiatives comprise all actions taken by local authorities and other official bodies. The analysis of 24 strategic documents prepared by the authorities of gminas (administrative units smaller than poviats) and cities that are part of the four poviats covering the region of Żuławy Wiślane (Fig. 2) has shown that authorities of 15 gminas and towns recognise cultural heritage as strengths when preparing SWOT analyses while the authorities of 12 analysed regions formulated strategic objectives for development, including protection, or exploiting cultural heritage. However, those objectives are very general and sweeping, for instance, "shaping identity of the city on the basis of its history" (Pruszcz Gdanski); "preserving landscape values" (Kolbudy); "tourist exploitation of cultural heritage" (Miłoradz); "promoting heritage and legacy left by the previous generation, both tangible and intangible" (Nowy Staw); "supporting initiatives aimed at preserving cultural heritage" (Nowy Dwór Gdański). What may be alarming is that in many documents cultural heritage is mentioned only in the context of exploiting it in order to develop the tourist industry. Obviously, tourism can contribute to the preservation of objects of historical values, but it can also pose a threat. Moreover, only in the case of one document, prepared by the authorities of Miłoradz, Żuławy Wiślanc was mentioned as a region and the gmina of Miłoradz as a part of it. It is clear that there is no general strategy for preserving the cultural landscape of Żuławy Wiślane as a whole. The authorities of the administrative units covering the region seem not to feel that their gimnas are part of a greater cultural whole. Such an approach does not guarantee proper sustainable development of the cultural landscape of Żuławy Wiślane.

The bottom-up initiatives include actions taken by NGOs, private associations and activists who are actively involved in promoting and preserving the cultural landscape of Żuławy Wiślane. Among the most active ones are:

- **Klub Nowodworski** [*Association of Nowy Dwor Gdanski Enthusiasts*] that manages the largest museum devoted to Mennonite culture in Żuławy Wiślane—the Żuławy Historical Park in Nowy Dwór Gdański organises many initiatives and events promoting Żuławy (e.g. "Żuławy. I like it!"; "Żuławy Initiative"; "International Meetings of Mennonites" and others) and supports the preservation of the cultural landscape of the region;
- **Stowarzyszenie "Kochamy Żuławy"** [*"We love Żuławy" Association*] organising trips and other events promoting the region among the locals and tourists;
- **Stowarzyszenie Żuławy Gdańskie** [*Żuławy Gdańskie Association*] focused on the eastern subregion of Żuławy Wiślane;
- **Żuławska Lokalna Grupa Działania** [*Żuławy Local Action Group*] associating local authorities and NGOs;
- **Lokalna Grupa Działania—Żuławy i Mierzeja** [*Local Action Group—Żuławy and the Spit*];
- **Tiegenhof—Kreis Großes Werder** [*Association of former residents of Nowy Dwór Gdański*];
- **Stowarzyszenie Żuławy** [*Żuławy Association*].

There are also private people, real enthusiasts of the region, taking actions aimed at preserving local cultural heritage: Marek Opitz, Łukasz Kępski, Mariola Mika, Marzena Bernacka-Basek, Artur Wasielewski, Dariusz Juszczak, Jerzy Domino, Roman Klim, Wojciech Hryniewiecki-Fiedorowicz, just to name a few.

Taking into consideration the scale of the bottom-up initiatives, the commitment of people involved in them and their knowledge as well as the general willingness of the local authorities to support such initiatives, it may be stated that there is a great chance that the cultural landscape of Żuławy Wiślane is to be developed in a sustainable way.

7 Summary

The landscape of Żuławy Wiślane is an excellent example illustrating the American approach to defining cultural landscapes. It is a region where natural and human-made elements are so tightly connected that it is even difficult to describe them separately. The element water which was the primary force shaping the landscape of the region has influenced human actions so deeply that it is present in almost every anthropogenic object visible in the landscape. What is more, the cultural landscape of Żuławy Wiślane has been created by numerous consecutive groups of people with different backgrounds and cultures, who were migrating there and brought their spiritual culture with them. Such richness of various cultural and religious influences has made the cultural landscape of Żuławy Wiślane unique, both in Poland and in Eastern Europe.

As the process of initiating and implementing changes into the landscape is conditioned environmentally and culturally and culture itself is now perceived as one of the sustainability pillars, the basic principles of sustainable development may apply to elements of cultural heritage as well. The cultural landscape of Żuławy Wiślane, which has been created by numerous generations with no living descendants, is the heritage of the past, just like non-renewable resources, so the principle of interregional and intraregional equity and the precautionary principle apply in this case.

The analysis of the cultural landscape of Żuławy Wiślane according to the categories of the STAMEX system has shown that it is possible to use methods of environmental assessment to describe cultural landscapes. The three categories constituting the system are allowed to describe the analysed region in a methodical and complex manner and to determine all elements of cultural heritage, their present condition and the initiatives aimed at their preservation and sustainable development.

Literature

Buttimer A (1978) *Charism and Context: the Challenge of La Geographie Humaine*. In: Ley D, Samuels MS (eds) Humanistic geography, prospects and problems, vol 15. Routlege, London and New York, pp 58–76

Butzer KW (1978) *Cultural perspectives on geographical space*. In: Butzer KW (ed) Dimensions of human geography: essays on some familiar and neglected themes. The University of Chicago, Department of Geography, Research Paper 186

Cernea M (2001) Cultural heritage and development. A framework for action in the Middle East and North Africa. World Bank, Washington

Convention on the protection and promotion of the diversity of cultural expressions. http://unesdoc.unesco.org/images/0014/001429/142919e.pdf. 3 July 2017

Culture and sustainable development: a framework for action (1999) World Bank, Washington

Czyżewska A, Kostarczyk A (1989) *Problematyka ochrony i kształtowania środowiska kulturowego Polski w planie przestrzennego zagospodarowania kraju*. In: Paterka T (ed) Przestrzeń kulturowa w planowaniu przestrzennym (discussion materials), Biuletyn KPZK PAN, 142, Warszawa

Dobrowolska M (1948) Dynamika krajobrazu kulturalnego. Przegląd Geograficzny **XXI**(3–4), 151–205

Eberhardt P (2015) Poglądy antropogeograficzne i geopolityczne Friedricha Ratzla. Przegląd Geograficzny 87(2):199–224

European Landscape Convention. https://rm.coe.int/1680080621. 1 July 2017

European Mennonite Network. http://www.eumen.net. 23 July 2017

Greffe X, Pflieger S, Noya A (2005) Culture and local development. OECD, Paris

Harding R, Fisher E (eds) (1995) Perspectives on the precautionary principle. The Federation Press, Sydney

Hawkes J (2001) The fourth pillar of sustainability—culture's essential role in public planning. Common Ground Publishing, Melbourne

Homans GC (1942) English villagers in the thirteenth century. Harvard University Press, Cambridge, Mass

Homans GC (1969) The explanation of English regional differences. Past and Present XLII:32–34

Janikowski R, Krzysztofek K (eds) (2009) *Kultura a zrównoważony rozwój. Środowisko, ład przestrzenny, dziedzictwo w świetle dokumentów UNESCO i innych organizacji międzynarodowych.* Polski Komitet ds. UNESCO, Warszawa

Kiełczewska-Zalewska M (1956) O powstaniu i przeobrażaniu kształtów wsi Pomorza Gdańskiego. Prace Geograficzne IG PAN 5:9–178

Kołodziejski J (1989) *Wprowadzenie.* In: Paterka T (eds) Przestrzeń kulturowa w planowaniu przestrzennym (discussion materials), Biuletyn KPZK PAN, 142, Warszawa

Kondracki J (2000) Regionalizacja fizycznogeograficzna Polski. PWN, Warszawa

Kowalczyk A (ed) (2008) Turystyka kulturowa. Spojrzenie geograficzne, Uniwersytet Warszawski, Wydział Geografii i Studiów Regionalnych, Warszawa

Lewis PF (1979) Axioms for reading the landscape. Some guides to the American scene. In: Meinig DW (ed) The interpretation of ordinary landscapes: geographical essays. Oxford University Press, New York, Oxford, pp 11–32

Lipińska B (2011) Żuławy Wiślane. Ochrona i kształtowanie zabytkowego krajobrazu. Stowarzyszenie Żuławy, Nowy Dwór Gdański

Medowski T (1989) Przesłanki ochrony i kształtowania środowiska kulturowego wsi. In: Paterka T (eds) Przestrzeń kulturowa w planowaniu przestrzennym (discussion materials), Biuletyn KPZK PAN, 142, Warszawa

Mężyński K (1960–1961) O mennonitach w Polsce. Rocznik Gdanski, XIX/XX:185–255

Mierzeja i Żuławy. http://mierzejaizulawy.pl. 28 July 2017

Mitchell N, Barrett B, Brown J (2014) Exploring Opportunities for Enhancing Relevancy and Sustainability through Cultural Landscape Conservation. In: Weber S (ed) Protected areas in a changing world: proceedings of the 2013 George Wright Society conference on parks, protected areas and cultural sites. George Write Society, Hancock, Michigan, pp 115–120

Murzyn-Kupisz M (2013) Kultura i dziedzictwo kulturowe a rozwój zrównoważony. In: Strzelecki Z, Legutko-Kobus P (eds) Gospodarka regionalna i lokalna a rozwój zrównoważony. Studia KPZK PAN, CLII, pp 92–105

Myga-Piątek U (2010) Transformation of cultural landscapes in the light of the idea of sustainable development. Problemy Ekorozwoju—Prob Sustain Dev 5(1):95–108

Navrud S, Ready RC (eds) (2002) Valuing cultural heritage. Applying environmental valuation techniques to historic buildings, monuments and artifacts. E. Elgar, Cheltenham

Opitz M (1998) *Żuławy – czas przełomu.* Nowy Dwór Gdański

Palmer R, Purchla J (eds) (2010) Kultura a rozwój 20 lat po upadku komunizmu w Europie. MCK, Kraków

Pascual J (2009) Polityki kulturalne, rozwój społeczny i innowacyjność instytucjonalna, czyli: dlaczego potrzebujemy Agendy 21 dla kultury? In: Kłosowski W (ed) Kierunek Kultura. Promocja regionu przez kulturę. Mazowieckie Centrum Kultury i Sztuki, Warszawa, pp 33–42

Pawłowski A (2009) *Rewolucja rozwoju zrównoważonego.* Problemy Ekorozwoju—Prob Sustain Dev 1(1):23–32

Ramsar sites information service. https://rsis.ramsar.org. 31 July 2017

Rembowska K (2002) Kultura w tradycji i we współczesnych badaniach geograficznych. Wydawnictwo Uniwersytetu Łódzkiego, Łódź

Solon J (2004) Ocena zrównoważonego krajobrazu. W poszukiwaniu nowych wskaźników. In: Kistowski M (ed) Studia ekologiczno-krajobrazowe w programowaniu rozwoju zrównoważonego. Przegląd polskich doświadczeń u progu integracji z Unią Europejską. Gdańsk, pp 49–58

Streeten P (2011) Culture and Economic Development. In: Ginsburgh VA, Throsby D (eds) Handbook of the economic of art and culture. Elsevier, North-Holland, Amsterdam

Strupińska K (ed) (2001) Żuławy Delty Wisły na przełomie tysiącleci: Żuławy Wiślane – unikalny obszar w Polsce i Europie, 1, Fundacja "Ecobaltic", Gdańsk
Szafran P (1981) Żuławy Gdańskie w XVII wieku. Wydawnictwo Morskie, Gdańsk
Szukalski J (1975) Żuławy Wiślane. WSiP, Warszawa
Throsby D (1995) Culture, economics, sustainability. J Cult Econ 19:199–206
Throsby D (2003) Cultural Sustainability. In: Towse R (ed) A handbook of cultural economics. E. Elgar, Cheltenham
UNESCO (1995) Our creative diversity. World commision on culture and development 1995. UNESCO, Paris
UNESCO (1998) Final report of intergovernmental conference on cultural policies for development: the power of culture 1998. UNESCO, Paris
UNESCO (2011) From green economies to green societies, UNESCO's commitment to sustainable development 2011. UNESCO, Paris
Throsby D (2010) Ekonomia i kultura. NCN, Warszawa
Wilczyński WJ (2005) Ewolucja poglądów geograficznych na środowisko. In: Podstawowe idee i koncepcje w geografii, Warszawa
Żuławy Info. www.zulawy.infopl.info. 7 July 2017

Application of Statistical Methods to Predict Beach Inundation at the Polish Baltic Sea Coast

P. Aniśkiewicz, P. Łonyszyn, K. Furmańczyk and P. Terefenko

Abstract In coastal zones, interactions between oceans and lands are very dynamic. Coastal and flood protection has a great importance for the safety of people. One of the approaches to predict beach inundation is to use simple empirical equations which have been successfully adopted by the coastal engineering branch. This research is focused on the application of wave run-up formulas (R_2) to predict beach inundation at the Polish Baltic Sea coast. Nine R_2 formulas were tested in six cross-sections at the beaches in Międzyzdroje, Ustronie Morskie, Sarbinowo, Darłówko, Lubiatowo and Dębki. Sea level elevation from tide-gauges and wave conditions from WAM model were used. The range of temporary seal level was from −0.29 to 0.39 m above mean sea level elevation defined as −0.08 m in Kronsztadt-86 Ordnance Datum. Maximum wave height was about 2.63 m and wave period was between 2.01 and 7.63 s. The results showed one extreme wave run-up (R_2) formula as percentage values which gave proper visual adjustment without overestimation and underestimation and statistically significant correlation (with 95% confidence level) in Sarbinowo (0.79), Lubiatowo (0.60) and Ustronie Morskie (0.49).

Keywords Beach inundation · Wave run-up · The Baltic Sea · Coast

1 Introduction

The climate on the Earth has been continually changing. However, during the last century increasing temperature trend is observed. The rate of annual warming for global land areas over the 1901–2000 period was estimated at 0.007 °C per year

P. Aniśkiewicz (✉)
Institute of Oceanology Polish Academy of Sciences, Sopot, Poland
e-mail: aniskiewicz@iopan.gda.pl

P. Aniśkiewicz
Centre for Polar Studies National Leading Research Center, Sosnowiec, Poland

P. Łonyszyn · K. Furmańczyk · P. Terefenko
University of Szczecin, Szczecin, Poland

T. Zielinski et al. (eds.), *Interdisciplinary Approaches for Sustainable Development Goals*, GeoPlanet: Earth and Planetary Sciences,
https://doi.org/10.1007/978-3-319-71788-3_7

(Jones and Moberg 2003). Rising global temperatures impact the stronger glacier and sea-ice melting and global sea-level rise (SLR). Scientists predict that in response to glacier melting and global warming, the sea level will rise up to 60 cm by the year 2100 (IPCC 2007). The reconstruction of global mean sea level from 1879 to 2004 showed a 20th century rate of SLR at about 1.7 ± 0.3 mm per year (Church and White 2006). Douglas (1991) estimated SLR for 91 years (1880–1980) as a 1.8 ± 0.1 mm per year from 21 stations from nine oceanic regions (North Sea, English Channel, Atlantic, Mediterranean, Pacific, North American West Coast, Central America, southeastern North America and northeastern North America). Global SLR calculated for 1993–1998 from Topex/Poseidon satellite was estimated as 3.2 ± 0.2 mm per year (Cabanes et al. 2001). Church and White (2011) calculated SLR at about 3.2 ± 0.4 mm per year from the 17-year (1993–2009) from satellite data and 2.8 ± 0.88 mm per year from the 130-year (1880–2009) sea level measurements. 20th century SLR is about 160 mm per year (Church and White 2006) and for the period between 1880 and 2009, it is about 210 mm (Church and White 2011). The long-term forecast of sea-level along the Polish Baltic Sea Coast between 1901 and 1990 was estimated as 0.012 ± 0.002 mm per year in Świnoujście and Kołobrzeg and 0.016 ± 0.002 in Gdańsk (Wróblewski 1994).

The consequences of SLR may be dangerous for people living on the Earth. McGranahan et al. (2007) calculated that 10% of world population live in low-elevation coastal areas. Whereas Rotnicki (1995) and Williamson (1992) described that almost 50% of the population living in all land areas settled down in zones around 50 km from coastal belts. That is the reason why coastal management policy and climate policy should be focused on effects in climate-induced SLR.

Regional meteorological effects like storm surges, long-term trends from land movement and ocean surface, and also modes of climate variability, have a big impact on coastal area environment (Church and White 2011). The direct effect of SLR is a saltwater intrusion into surface water and an inundation of coastal areas. The long-term sea level rise may amplify erosion and a saltwater intrusion to ground waters (Nicholls and Cazenave 2010; Nicholls et al. 2007).

Polish coastal zone with a plenty of fauna and flora is a specific terrestrial tidal and shelf ecosystem. It may be described as an area with water depth less than 20 m. Almost 25% of the whole Baltic Sea is shallower than this depth. Likewise, islands (except few of them in central basins) are a part of the coastal zone (Leppäranta and Myrberg 2009; following Leppäranta 1981). In the Polish Baltic Sea coast, the weather is strongly influenced by westerly polar maritime air masses. Less impact on the weather has arctic air from the Norwegian Sea and polar-continental air masses from the Eastern Europe. The precipitation mostly occurs in summer and autumn—about 80% of the annual value. The strongest wind in the coastal zone is observed during the autumn-winter season and the lowest between May and August. The frequency of stormy weather ranges at the Polish Baltic Sea coast between 2 and 5%. The difference depends on season and area.

This stormy weather causes storm surges at the coast. Maximum sea level value at the Polish Baltic Sea coast is caused by east-northerly and northerly winds (Zeidler et al. 1995). Stormy weather has a big impact on environmental transformation at beaches, which may be classified as a contact zone between sea and land, where intensive water masses and energy exchange occurs (Rotnicki 1995). It may be dangerous for tourist and resident people living in coastal municipalities.

The national "Program of the coastal protection for the years 2003–2023" started in the year 2004. It was the conclusion of the 5 years of work. This strategy introduces the following activities:

- levees,
- seawalls,
- groynes,
- breakwaters,
- artificial reefs,
- beach nourishment.

All of them depend on the needs at the concrete place. From six locations used in the research three of them could have an important influence on the accuracy of the predictions. In these circumstances, the achieved results should be much appreciated. Regarding the financial side of the Strategy there are planned the following expenses in four out of six locations from this research:

- in Sarbinowo on the beach length of 4.2 km—20,000 PLN,
- in Darłówko on the beach length of 0.6 km—50,000 PLN,
- in Ustronie Morskie on the beach length of 5.9 km—40,000 PLN,
- in Międzyzdroje on the beach length of 1.7 km—10,000 PLN.

The planned actions were realized only partly till the end of 2013. In Sarb, inowo, in the year 2012, the local concrete seawall and beach nourishment was modernized. In Ustronie Morskie, since 2005, a few nourishments and new groynes were developed. In Darłówko, since 2011, there was a modernization of breakwater and building of the artificial reef. Only in Międzyzdroje, nothing has been done until the end of 2013 (Łabuz 2013).

In Poland, prior to an adoption of the Strategy, the sediment from dredging was putting dumping sites to sea. In the years 1990–2008 only 43% of the dredging sediment nourished polish beaches (Staniszewska et al. 2014). At present, the proportion is different and many polish beaches systematically get the sand replenish. The actions are an effect of combining two aims. First, coastal protection by the movement of the coastline far from dunes or cliffs. Second, the augmentation of the beach capacity that gives better conditions for tourists. Some of the sediments come from works on waterways dredging but the demand is so big that the sediments are also taken from the Baltic seabed (Staniszewska et al. 2014).

The beach nourishment has an important impact on its shape and size—key factors of precise predictions of the wave run-up. Unfortunately, in tender documents (the ToR) of the maritime offices regarding the beach nourishment, there are no detailed guidelines about those two factors so it is impossible to adjust statistic formulas to increase the prediction accuracy. Moreover, the erosion of the 'new' sand is faster than the 'old'. It makes the shape of the beach changing very dynamically after the nourishment. All of these circumstances could strongly influence the results of predictions from statistical formulas. Of course, the way to limit the erosion of replenished sand is building coastal protecting constructions, especially groynes. Maritime Offices in Poland during the years were very prudent about this type of installations. It was quite recent when this attitude has changed and wooden groynes appeared for example in Ustronie Morskie. Since this construction dissipates the sea wave energy, it helped to rebuild the beach in this village.

Nowadays, scientists have been working on the development of flood prediction and coastal protection. One approach is to use models to predict, for example, sea water level, wave propagation, sediment transport and morphological changes of beaches. The second approach described in this paper is the calculation of maximum wave run-up (R_2). R_2 may be described as a level of run-up vertically measured depends on still water level (van der Meer 1998). Wave run-up is a very important indicator for building infrastructure in coastal zone where flood prediction is one of the main issues for the safety of people. In previous papers, many authors have been adapted wave run-up formulas to the ocean and marine coasts (e.g. Díaz-Sánchez et al. 2014; Hedges and Mase 2004; Nielsen and Hanslow 1991; Ruggiero et al. 1996; Mase 1989; Vousdoukas et al. 2012). Most of these equations are based on the Iribarren number created by Iribarren and Nogales (1949), where H is wave height and L_0 means the linear theory deep-water wave length (1), α is the slope angle (2) and T is regular wave period (3).

$$\xi_0 = \frac{\beta}{\sqrt{\frac{H}{L_0}}} \tag{1}$$

$$\beta = \tan(\alpha) \tag{2}$$

$$L_0 = \frac{gT^2}{2\pi} \tag{3}$$

Galvin (1972) modified this formula and wave height (H) component was replied to deep-water wave height (H_0) (4):

$$\xi_0 = \frac{\beta}{\sqrt{\frac{H_0}{L_0}}} \tag{4}$$

The Iribarren parameter (4) depends on types of wave break (Table 1) (Battjes 1974).

Table 1 Types of wave break depending on the Iribarren parameter

ξ_0	Type of wave break
$\xi_0 < 0.5$	Spilling
$0.5 < \xi_0 < 3.3$	Plunging
$\xi_0 > 3.3$	Surging and collapsing

Mase (1989) described wave-run formula based on the Iribarren number:

$$\frac{R_2}{H_0} = a \cdot {\xi_0}^b \tag{5}$$

where a and b are empirical parameters. He suggested the following values: 1.86 for a and 0.71 for b parameter.

Wave run-up formulas were adapted before to some parts of Polish Baltic Sea coast. Paprotny et al. (2014) recalibrated empirical parameters of Mase (1989) equation for hydrological conditions (5) in two test sites in Dziwnówek and Międzyzdroje using data from the year 2013. The results are: a = 1.29 and b = 0.72. Aniśkiewicz et al. (2016) validated this formula for data from another period between 1st June and 31st August 2014 in Dziwnówek. They have also validated other R_2 formulas (Guza and Thornton 1984; Díaz-Sánchez et al. 2014; Hedges and Mase 2004; Holman and Sallenger 1985; Nielsen and Hanslow 1991; Ruggiero et al. 1996; Stockdon et al. 2006; Vousdoukas et al. 2012; Mase 1989) in this region and modified them by adding temporary sea level value to all of them. That was done because most of the formulas were calibrated in laboratory conditions and they could be incorrect for natural beaches where sea level changes are a very important factor.

The results obtained by the previous authors showed that the verified statistical formulas are not universal and cannot be easily adapted for predicting beach inundation in the Polish coastline. The values of the correlations between real and predicted data were in the range of 0.45 and 0.53 (Aniskiewicz et al. 2016).

That was the starting point to the analysis made by authors of this paper. The idea was to verify formulas on other places in the Polish coastline to exclude the possibility that the uniqueness of the place, in this case Dziwnówek, has a strong impact on the achieved results quality. This one village could not be representative for all coastline of the country.

2 Research Area

That is the reason why we have decided to choose other locations along the polish coastline and analyzed wave run-up formulas in this area. Another six profiles have been chosen. Listing from west to east, there were beaches in Międzyzdroje, Ustronie Morskie, Sarbinowo, Darłówko, Lubiatowo and Dębki (Fig. 1). The first one is the closest one to Dziwnówek but still, it is 23 km away.

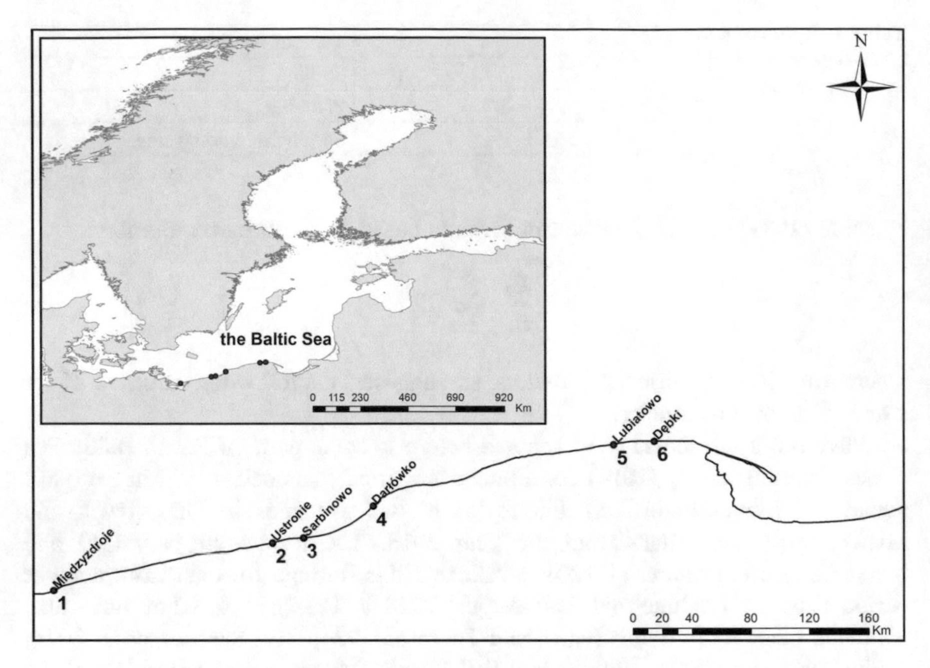

Fig. 1 The location of municipalities (based on www.arcgis.com)

Darłówko is a seaside village in north-western Poland within West Pomeranian Voivodeship. It is considered as an unofficial district of Darłowo. At present, the town's population is over 14,000. It is a very popular tourist destination during a summer season with very crowded beaches. The width of the beach in Darłówko is 41 m at the place where the profile has been taken. At the mouth of Wieprza, the river that crosses the town, there is a little harbor with a long breakwater at the entry. Another example of invasive policy regarding coastal protection is an artificial reef that protects the local beach from erosion. This construction is visible at the web camera whose signal has been used during the research. It has been taken into consideration that this type of construction could have a strong impact on estimations of the wave run-up using statistical formulas.

Dębki is the smallest village in this research. Its population is about 160. It is located within Pomeranian Voivodeship in northern Poland at the mouth of Piaśnica river. Historically this little village was an important place as the river crossing Dębki was a border between Poland and Germany during the interwar period. Presently it is a tourist summer destination. Here the width of the beach was the largest of all six locations in this research. It was 45 m at the place of the profile.

In Lubiatowo the profiles for the research were taken at the coastline in the place that is not located in Lubiatowo, as the village is 2.5 km deep in the land, but this hamlet is the nearest one. It is located within Pomeranian Voivodeship in northern Poland and its population is about 180. The local sandy beaches are very popular among caravanning fans as in local forests there are a lot of camping camps. At the place where the profile was taken, the width of the beach was only 27 m.

Międzyzdroje is a seaside town in north-western Poland within West Pomeranian Voivodeship. The whole town is located on the biggest Polish island—Wolin. It is one of the most important and popular polish tourist resorts although its population is only 5500. It has one of the longest seaside piers in Europe—almost 400 m long. The coast in Międzyzdroje is a mix of dunes and high cliffs with sandy beaches 35 m wide in measured profile. Another interesting thing about the local coast is a shoal that periodically emerges from the water. This phenomenon is an important factor regarding the analysis of the wave run-up as it could strongly distort the results.

Sarbinowo is a small village in north-western Poland within West Pomeranian Voivodeship. The population is less than 600. The width of the beach at the place where the profiles have been taken is 30 m. The village has no major attractions. There is one of the oldest constructions and it is a retaining wall protecting local small cliffs from sloping down and erosion. The local seaside promenade in Sarbinowo is, in fact, a concrete seawall protecting residential and tourism housing that is now in a direct vicinity of the beach as an effect of coastal erosion.

Ustronie Morskie is a popular tourist village at the Polish seaside in north-western Poland within West Pomeranian Voivodeship. The population of the village is about 2000. The profile has been taken at the place where the beach is very narrow—only 15 m. In the late 90s there was built a concrete seawall. The primary effect was very unexpected as the beach disappeared. The present width of the beach is a result of sand replenishing and adding a few wooden groynes. That process was started in the year 2005 and is still in progress. It has to be taken into consideration talking about the accuracy of the predictions.

3 Methodology

In this paper, well-known wave run-up formulas were used to predict beach inundation in six profiles along the Polish Baltic Sea coast. How were these profiles chosen? The first step was to search the website with available cameras live view. The cameras were selected if they were pointing on the beach in the angle that would make possible an observation of wave run-up phenomena. Some of the cameras along the coast were moving, mostly horizontally, so the final step was to choose the static ones. It was important because only these cameras allowed to save photos always in the same place. Only the six website cameras fulfilled the mentioned criteria. Data from cameras were collected from the period between 1st April and 31st July 2015 two times per day, depending on their availability. For all cameras these profiles were chosen which were visible from them. Cross-sections are located at about 412.7 km in Międzyzdroje, 319.7 km in Ustronie Morskie, 306.7 km in Sarbinowo, 270.2 km in Darłówko, 163.7 km in Lubiatowo and 147.5 km in Dębki. These values relate to the Maritime Office's kilometrage.

Profiles were done using GPS RTK. Measuring sandy coast by this equipment is a common approach to monitor morphological changes in a short-time scale (e.g. Bugajny et al. 2013, 2015; Ortega-Sánchez et al. 2008). The horizontal resolution of GPS-RTK used in this research is about ± 1.5 cm and vertical at around ± 2 cm (Bugajny and Furmańczyk 2014). XY coordinates were measured in every 1 m along the profile lines from the mean sea level (MSL) point to dune base (in Darłówko, Dębki, Lubiatowo, Międzyzdroje and Sarbinowo) and to the concrete seawall in Ustronie Morskie. MSL point is defined as a −0.08 m in Kronsztadt-86 Ordnance Datum. During the every single XY measurement the photos from the camera were saved. This allowed to recognize the range of wave run-up with 1 m accuracy. The MSL point was 0% beach inundation point on the video camera. 100% flood means that extreme wave reached the dune base. Thanks to this approach the distance between every measurement on the profiles could be easily transformed to distance in pixels. Detailed information about this part of methodology was described in Paprotny et al. (2014).

To define wave break type we had to calculate the Iribarren number (4) for every cross-section. The length and the width of the beach were measured from Digital Elevation Model prepared in the year 2013 by the Central Geodetic and Cartographic Documentation Center (CODGIK). Peak wave period and deep-water wave height were provided by the Ocean Wave Prediction Model (WAM) adapted to the Baltic Sea by the Interdisciplinary Centre for Mathematical and Computational Modeling (Cieślikiewicz and Paplińska-Swerpel 2008). The authors used data from WAM point no. 12 (Międzyzdroje), no. 66 (Ustronie Morskie), no. 115 (Sarbinowo), no. 240 (Darłówko), no. 647 (Lubiatowo) and no. 649 (Dębki). Based on these data the Iribarren numbers were calculated.

Almost all wave run-up formulas except one, created by Guza and Thornton, (1984) are based on the Iribarren number (Table 2). In two examples the authors omitted it when the value of the Iribarren number was not significant. Nielsen and Hanslow (1991) neglected it when beach steepness was higher than 0.10. They were also the only scientists who included in their formulas the mean sea level (SWL). This value was replaced by temporary sea level by Aniśkiewicz et al. (2016). Most of the formulas used one constant but there were some with two and Vousdoukas et al. (2012) proposed even three constants.

Based on types of wave break the proper formula to calculate R_2 was chosen. The decision was to compare modified formulas (Table 3) with real beach inundation data (BI_R). Except 9th equation in Table 3 which was made by Paprotny et al. (2014) all of them were modified by Aniśkiewicz et al. (2016) and tested in another area at the Polish Baltic Sea coast. The range of applicability was the same for modified formulas as it was for based equations (Table 2). The results showed also better prediction of beach inundation with additional sea level (BI_{MS}) (Aniśkiewicz et al. 2016). Sea level variability was provided by the Institute of Meteorology and Water Management (IMGW) tide-gauge stations: Świnoujście (for Międzyzdroje), Kołobrzeg (for Ustronie Morskie, Sarbinowo and Darłówko) and Łeba (for Lubiatowo and Dębki). The range of wave height, peak wave period and temporary sea level in every profile are shown in the Table 4.

Table 2 Wave run-up formulas with a range of applicability

Lp	Original formula	Range of applicability	Authors
1	$R_2 = 1.20 \cdot \xi_0 \cdot H_0$	$\xi_0 > 0.3$	Díaz-Sánchez et al. (2014)
	$R_2 = 0.32 \cdot H_0$	$\xi_0 < 0.3$	
2	$R_2 = 0.55 \cdot H_0$	–	Guza and Thornton (1984)
3	$R_2 = (0.34 + 1.15 \cdot \xi_0) \cdot H_0$	$\xi_0 < 2.4$	Hedges and Mase (2004)
4	$R_2 = (0.2 + 0.83 \cdot \xi_0) \cdot H_0$	$\xi_0 > 0.5$	Holman and Sallenger (1985)
5	$R_2 = H_0 1.86 \cdot {\xi_0}^{0.71}$	$1/30 < \beta < 1/5$ $H_0/L_0 \geq 0.007$	Mase (1989)
6	$R_2 = (SWL + 1.19 \cdot \xi_0) \cdot H_0$	$\beta < 0.10$ or $\Omega > 6$	Nielsen and Hanslow (1991)
	$R_2 = (SWL + 0.1 \cdot \sqrt{H_0 L_0}) \cdot H_0$	$\beta > 0.10$ or $\Omega < 6$	
7	$R_2 = 1.29 \cdot H_0 \cdot {\xi_0}^{0.72}$	–	Paprotny et al. (2014)
8	$R_2 = 0.27 \cdot \frac{\xi_0}{\sqrt{\beta}} \cdot H_0$	$\xi_0 > 0.5$	Ruggiero et al. (1996)
	$R_2 = 0.5 \cdot H_0 - 0.22$	$\xi_0 < 0.5$	
9	$R_2 = 0.53\beta\sqrt{(H_0 L_0)} + 0.58\xi_0\sqrt{\frac{H_0^3}{L_0}} + 0.45$	–	Vousdoukas et al. (2012)

Table 3 Modified wave run-up formulas

Lp	Modified formula	Authors
1	$R_2 = 1.20 \cdot \xi_0 \cdot H_0 + S$	Aniśkiewicz et al. (2016)
	$R_2 = 0.32 \cdot H_0 + S$	Aniśkiewicz et al. (2016)
2	$R_2 = 0.55 \cdot H_0 + S$	Aniśkiewicz et al. (2016)
3	$R_2 = (0.34 + 1.15 \cdot \xi_0) \cdot H_0 + S$	Aniśkiewicz et al. (2016)
4	$R_2 = (0.2 + 0.83 \cdot \xi_0) \cdot H_0 + S$	Aniśkiewicz et al. (2016)
5	$R_2 = H_0 1.86 \cdot {\xi_0}^{0.71} + S$	Aniśkiewicz et al. (2016)
6	$R_2 = (S + 1.19 \cdot \xi_0) \cdot H_0$	Aniśkiewicz et al. (2016)
	$R_2 = (SWL + 0.1 \cdot \sqrt{H_0 L_0}) \cdot H_0 + S$	Aniśkiewicz et al. (2016)
7	$R_2 = 1.29 \cdot H_0 \cdot {\xi_0}^{0.72} + S$	Paprotny et al. (2014)
8	$R_2 = 0.27 \cdot \frac{\xi_0}{\sqrt{\beta}} \cdot H_0 + S$	Aniśkiewicz et al. (2016)
	$R_2 = 0.5 \cdot H_0 - 0.22 + S$	Aniśkiewicz et al. (2016)
9	$R_2 = 0.53\beta\sqrt{(H_0 L_0)} + 0.58\xi_0\sqrt{\frac{H_0^3}{L_0}} + 0.45 + S$	Aniśkiewicz et al. (2016)

All results of R_2 were shown as a percentage value of modelled beach inundation (BI_M), where *h* is baseline height of the beach (6) (following Paprotny et al. 2014; Aniśkiewicz et al. 2016).

$$BI_M = \frac{R_2}{h} \cdot 100\% \quad (6)$$

The dependencies between h, SWL, S, R_2, dune base line and BI_M were presented in Fig. 2.

To compare modelled and measured data correlation coefficients (r) with 95% confidence level and statistical error (STE) for all formulas and cross-sections were calculated. STE is a standard deviation of differences between real and predicted data. It allows to determine how much the average value of measurements differentiates from its real value. In addition, systematic errors (SYE) to verify the quality of the predictions were computed. SYE represents environmental, human or tool calibration factors which may be difficult to detect and cannot be analyzed

Table 4 The range of input data for prediction of beach inundation

Lp	Name	Wave height [m]	Wave period [s]	Temporary sea level [m]
1	Międzyzdroje	0.01–1.35	2.01–6.30	−0.29–0.36
2	Ustronie Morskie	0.05–2.17	2.67–6.93	−0.24–0.39
3	Sarbinowo	0.05–2.17	2.67–6.93	−0.24–0.39
4	Darłówko	0.07–2.51	2.94–6.93	−0.24–0.39
5	Lubiatowo	0.05–2.63	2.43–7.63	−0.12–0.26
6	Dębki	0.05–2.63	2.43–7.63	−0.12–0.26

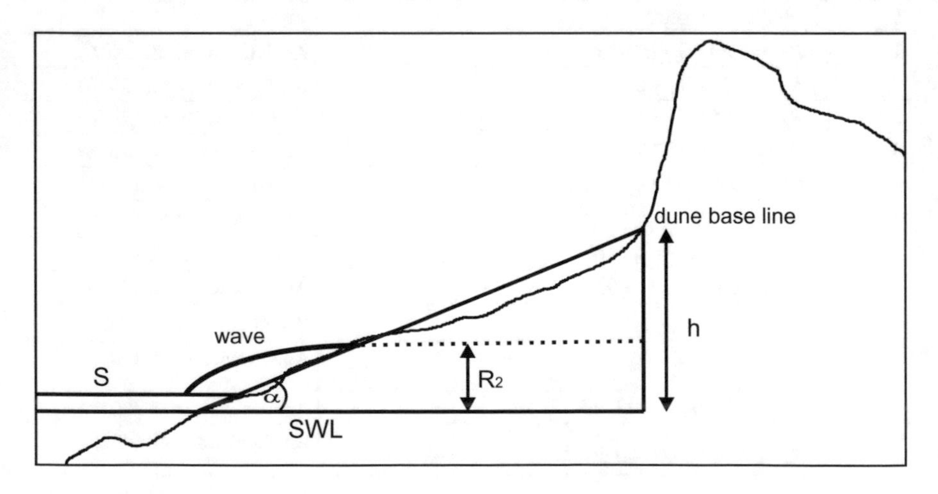

Fig. 2 The scheme of calculating R_2

statistically. Because it quantifies overestimation and underestimation of beach inundation the decision was to correct BI_M formulas by SYE (7).

$$BI_{MS} = \frac{R_2}{h} \cdot 100\% + SYE \tag{7}$$

The last part of the research was to select the ten highest values from real data to compare real extreme wave run-up (BI_R) with predictions (BI_{MS}). In case if the tenth highest value was observed in more than one control point, each such value was classified as the joint-tenth-best.

4 Results

The first step of this research was to calculate types of wave break for every profile. The length of the beach (L) was measured from the dune base to the mean sea level point. The shortest beaches (≤ 30 m) were observed in Ustronie Morskie (15 m), Lubiatowo (27 m) and Sarbinowo (30 m). In other cities the cross-sections were longer: 35 m (Międzyzdroje), 41 m (Darłówko) and 45 m (Dębki).

The calculation of slope angle (α) provided data that the biggest value occurred in Dębki (5.08°) and the lowest in Lubiatowo (2.12°). In other locations, it was 3.27° in Międzyzdroje, 3.35° in Darłówko, 4.5° in Sarbinowo and 4.95° in Ustronie Morskie (Table 5). This calculation allowed to find the Iribarren number (5) for all profiles. It ranged between 0.5 and 0.7. The authors decided to choose these wave run-up formulas which fulfilled the presumption $\xi_0 > 0.5$. Finally, nine equations were chosen and their predictions (BI_M) were correlated with real data (BI_R) (Table 6).

Only in three profiles (Lubiatowo, Sarbinowo and Ustronie Morskie) correlations between real and predicted values were statistically significant at 95% confidence level (Table 6). The cross-sections without significant correlation were rejected. In selected profiles r values were between 0.49 and 0.79. The highest value was calculated for the 8th formula in Sarbinowo and the lowest for equation no. 5 in Ustronie Morskie. The mean correlation for all profiles is higher than 0.5: 0.77 (Sarbinowo), 0.57 (Lubiatowo) and 0.55 (Ustronie Morskie). The correlation coefficient quantifies a relationship between observed and expected values but it

Table 5 Length of the beach (L), baseline height of the beach (h) and slope angle (α) calculated for every profile

Lp	Location	L [m]	h [m]	α [°]	ξ_0
1	Międzyzdroje	35	2.0	3.27	0.6
2	Ustronie Morskie	15	1.3	4.95	0.7
3	Sarbinowo	30	2.4	4.57	0.6
4	Darłówko	41	2.4	3.35	0.5
5	Lubiatowo	27	1.0	2.12	0.5
6	Dębki	45	4.0	5.08	0.7

Table 6 Correlation coefficients calculated for selected formulas

No.	Equation	Darłówko	Dębki	Lubiatowo	Międzyzdroje	Sarbinowo	Ustronie Morskie
1	$R_2 = 1.20 \cdot \xi_0 \cdot H_0 + S$	0.01	0.17	0.57	0.13	0.77	0.55
2	$R_2 = 0.55 \cdot H_0 + S$	−0.01	0.14	0.59	0.10	0.78	0.53
3	$R_2 = (0.34 + 1.15 \cdot \xi_0) \cdot H_0 + S$	0.02	0.13	0.54	0.14	0.73	0.54
4	$R_2 = (0.2 + 0.83 \cdot \xi_0) \cdot H_0 + S$	0.02	0.17	0.57	0.12	0.77	0.55
5	$R_2 = H_0 1.86 \cdot {\xi_0}^{0.71}$	0.03	0.2	0.50	0.14	0.75	0.56
6	$R_2 = (S + 1.19 \cdot \xi_0) \cdot H_0$	0.00	0.17	0.57	0.12	0.77	0.55
7	$R_2 = 1.29 \cdot H_0 \cdot {\xi_0}^{0.72} + S$	−0.04	0.12	0.61	0.10	0.79	0.53
8	$R_2 = 0.27 \cdot \frac{\xi_0}{\sqrt{\beta}} \cdot H_0 + S$	−0.01	0.13	0.58	0.09	0.78	0.52
9	$R_2 = 0.53\beta\sqrt{(H_0 L_0)} + 0.58\xi_0\sqrt{\frac{H_0^3}{L_0}} + 0.45 + S$	−0.09	0.07	0.60	0.07	0.79	0.49

Table 7 Statistical (STE) and systematic errors (SYE) calculated for all formulas and all profiles

No.	Sarbinowo	Lubiatowo	Ustronie Morskie	Sarbinowo	Lubiatowo	Ustronie Morskie
	STE [%]	STE [%]	STE [%]	SYE [%]	SYE [%]	SYE [%]
1	8.03	16.94	19.76	−0.12	−17.09	−19.65
2	8.17	21.66	19.57	2.61	−34.17	−10.84
3	12.08	21.44	23.87	−9.17	−40.00	−33.11
4	8.28	19.05	20.05	−0.16	−25.27	−18.43
5	15.46	21.66	24.09	−15.96	−44.59	−41.53
6	9.48	19.90	23.25	−4.44	−52.43	−24.86
7	10.12	19.58	22.56	−5.25	−30.43	−27.52
8	7.76	23.50	18.99	13.22	−10.34	8.56
9	7.48	15.10	17.76	−9.78	−49.72	−36.67

does not show model overestimation and underestimation. For example, the correlation may be high even if the predicted values are overestimated. That is why the statistical (STE) and systematic (SYE) errors were calculated. STE values were from 7.48% for 9th formula in Sarbinowo to 24.09% for equation no. 5 in Ustronie Morskie (Table 7). Systematic errors were high, up to 41.54% for 5th formula in Ustronie Morskie which means that wave run-up equations overestimated predictions. Adding SYE to BI formula (6) did not change correlation coefficients and the statistical errors but increased adjustment to real data.

In Sarbinowo, which is located in the middle part of the Polish Baltic Sea coast, the highest correlation was calculated for the 7th and the 9th formulas (0.79) and the lowest for the 3rd (0.73). The STE was the lowest for the formula no. 9 (7.48%) and the highest for the 5th (15.46%). Only for two equations (2nd and 8th) wave run-up was underestimated by 2.61% for the 2nd and 13.22% for the 8th (Table 7). In other cases the results were overestimated. The value of it was between 0.12% (1st Eq.) and 15.96% (5th Eq.). For 3rd and 5th formulas, extreme values were almost four times overestimated and correction with systematic error included did not change the results at all. The lowest visual overestimation of extreme values was observed in the 3rd, 5th and 6th equations (with systematic error correction) (Fig. 3). In this part of the research area, 11 control points were chosen with the real beach inundation value higher than 36% (Fig. 4). For two control points (10th and 11th) the values of BI_R were the same (46%). The best visual adjustment was for the 8th and the 9th formulas. For the equation number 5, the values were much over and underestimated depending on the value of real data.

For the cross-section in Lubiatowo the lowest correlation occurred for the 5th formula (0.50) and the highest for the equation no. 7 (0.61) (Table 6). The values of statistical errors were between 15.10% (9th Eq.) and 23.5% (8th Eq.). The range of SYE was between 10.34% for the equation no. 8 to 52.43% for the 6th formula (Table 7). The best visual adjustment with the SYE was for the 6th and the 9th equations (Fig. 5). Other formulas overestimated extreme wave run-up even after

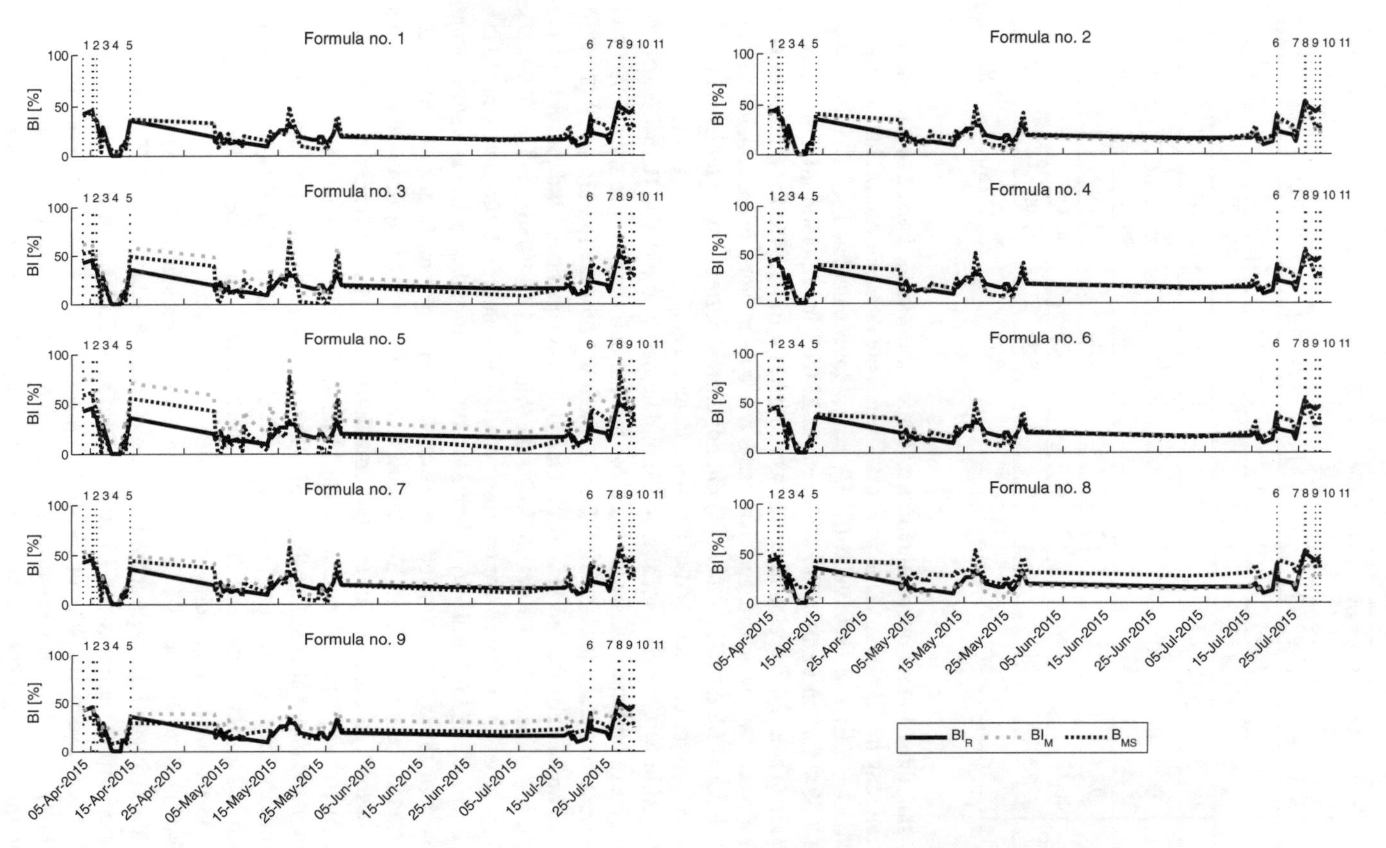

Fig. 3 The comparison of real beach inundation (BIR) with modelled (BIM) and modelled with SYE (BIMS) data with markers 11 control points

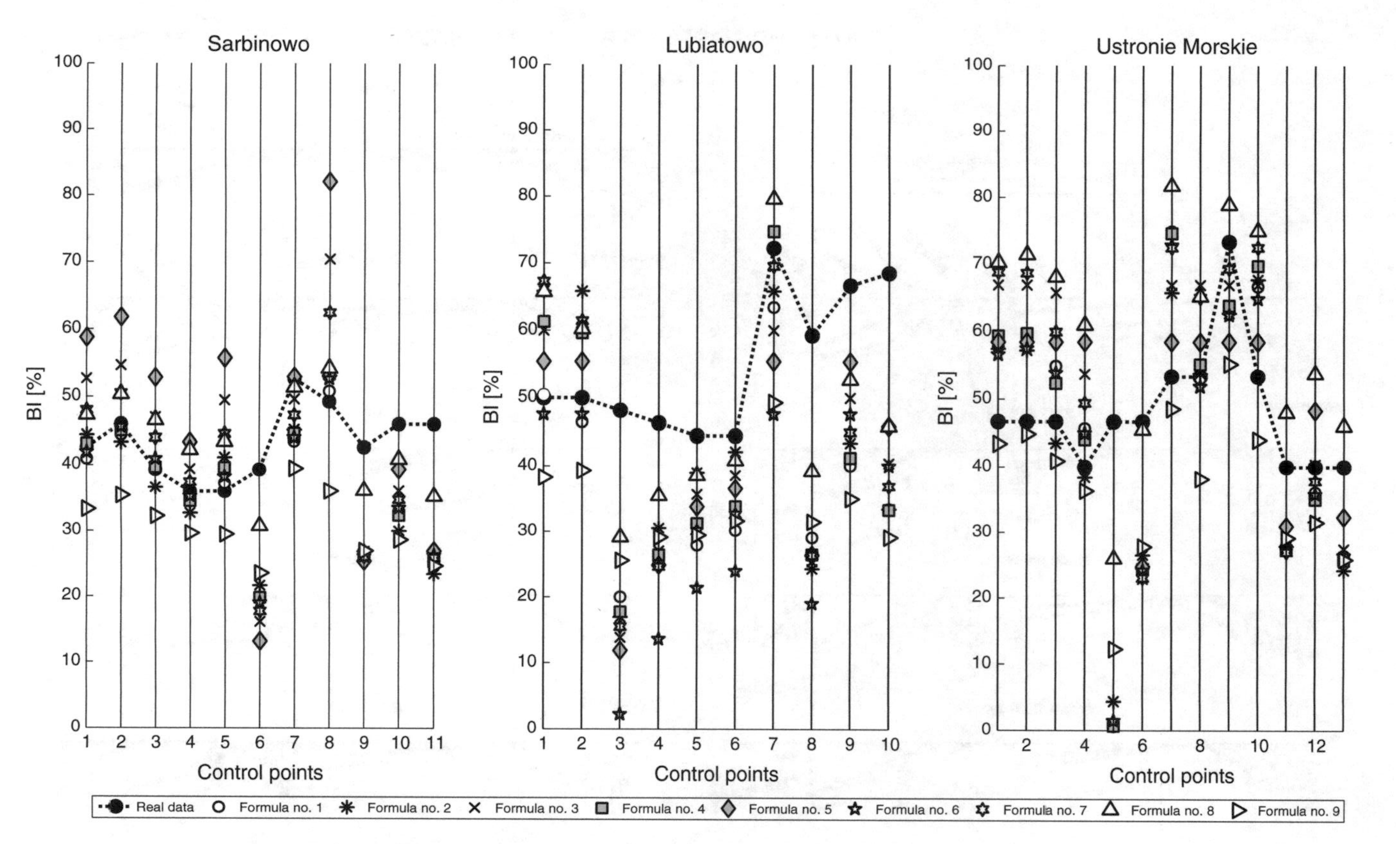

Fig. 4 The comparison of BIR and BIMS in control points

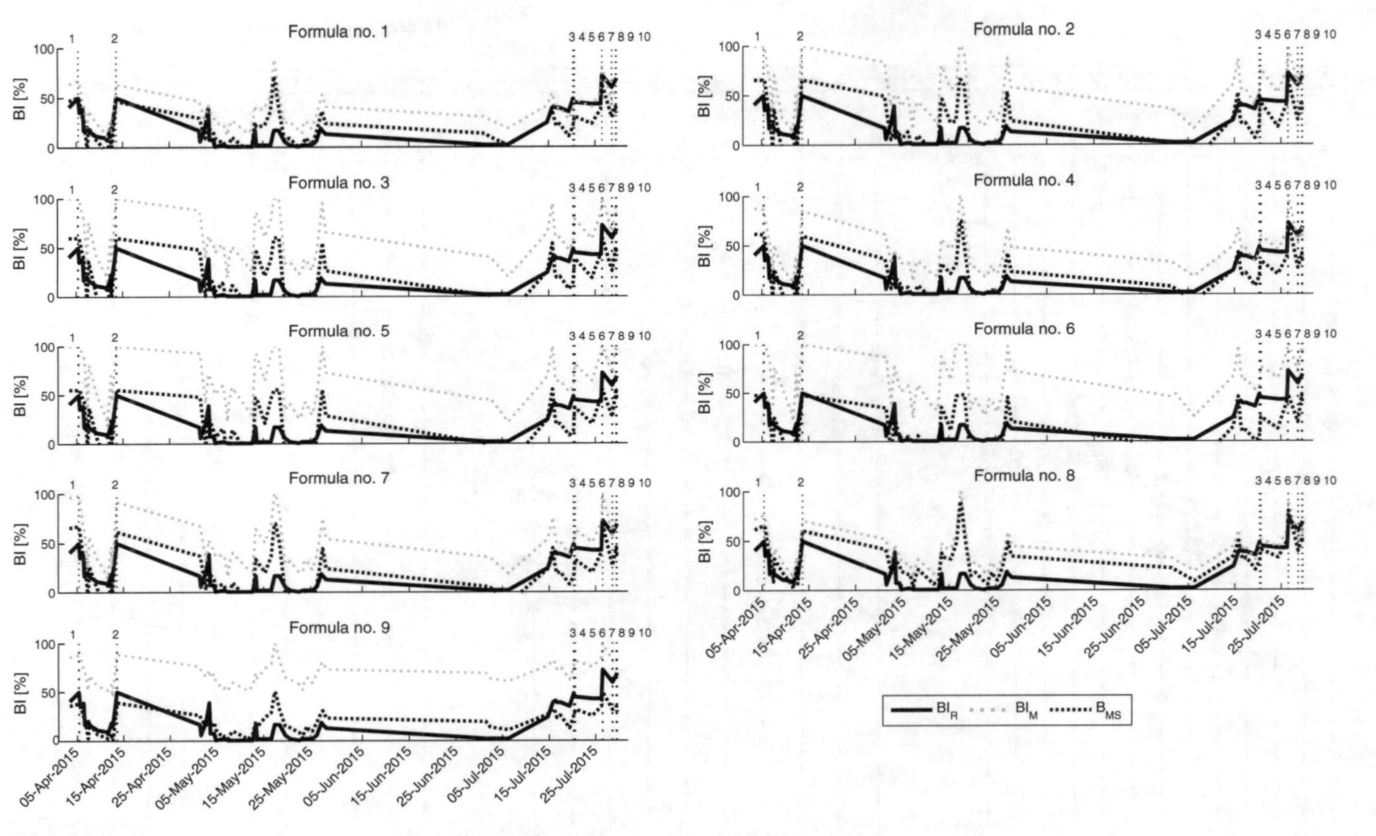

Fig. 5 The comparison of BIR with BIM and BIMS in Lubiatowo with marked 10 control points

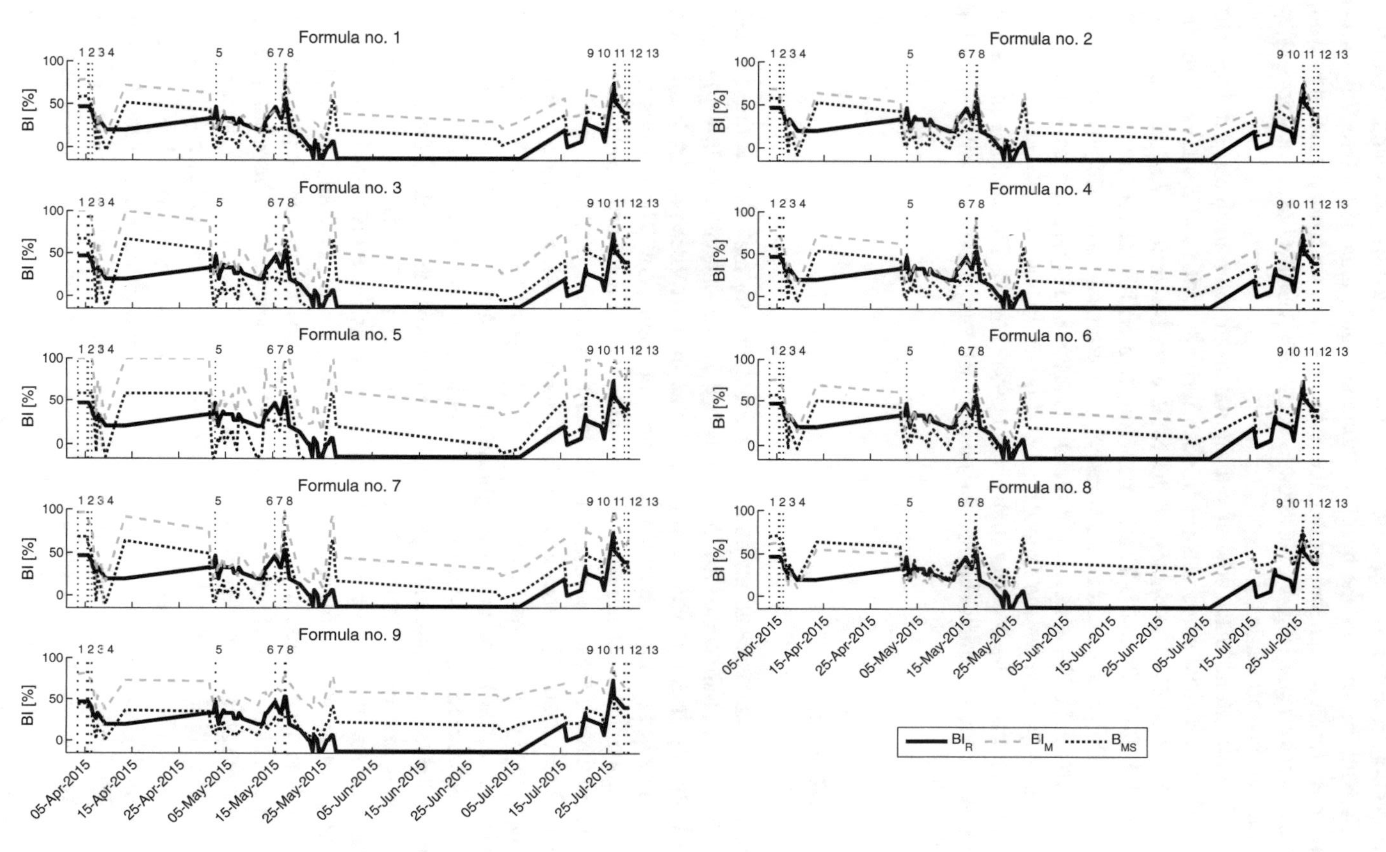

Fig. 6 The comparison of BIR with BIM and BIMS in Ustronie Morskie with marked 13 control points

the correction by systematic error. In this cross-section, ten BI_R were chosen for the analysis with the value of the beach inundation higher than 43%. Because correlations were lower than in Sarbinowo, the extreme values were not predicted as well as in the previous area (Fig. 4). Nevertheless, the best visual adjustment was easy to find for the 9th formula.

In Ustronie Morskie the lowest correlation from three analyzed profiles was observed. The highest value was for the 5th formula (0.56) and the lowest for the 9th (0.49) (Table 6). The values of the STE were between 17.76 (9th Eq.) and 24.09% (5th Eq.) (Table 7). The biggest negative SYE was for the 5th formula (−41.53%). Only for the equation number 8, the systematic error had positive value (8.56%) (Table 7). The correction of the BI_M for the SYE was a proper solution for this profile, especially for the formula no. 9 (Fig. 6). In this case, 13 control points were chosen where the BI_R was higher than 39% (Fig. 4). The analysis of control points in this cross-section showed the best adjustment of extreme values also for 9th formula.

5 Conclusions

The best correlations for all formulas were calculated for the cross-section located in Sarbinowo. In Lubiatowo and Ustronie Morskie, correlations were lower but still statistically significant at 95% confidence level. In other three profiles (Darłówko, Dębki and Międzyzdroje) correlations were not significant. The reasons of such situation are different. In the case of Darłówko, around the place where the profile was taken, there is located an artificial reef for the protection of local beaches from erosion. This construction may change hydrological conditions of water masses like a natural type of wave break, wave height or temporary sea level. In the case of Międzyzdroje, the 400 m long pier significantly influences the values of inundations and biases the predictions. Much more difficult is to explain the case of Dębki. Most probably the lack of statistical significance is a result of the mouth of the Przaśnica river that is nearby. The above-mentioned examples prove that wave run-up formulas may not always be correct for each profile at the Polish Baltic Sea coast. Statistical errors were no higher than 25% which may be acceptable, especially for simple equations. Because wave run-up formulas were created for the prediction of beach inundation, the one, giving good correlation and visual adjustment without high overestimation or underestimation, had to be selected. As it has been written before, the analysis should not be based only on a correlation coefficient but also on other factors like statistical and systematic errors and good visual adjustment. Despite the highest correlation calculated for the 7th formula, the best results were for the 9th formula. Comparing to the 7th equation, the difference in correlation coefficient was very low but statistical errors for every profile were the lowest. In every other equation, there was a problem with the prediction of extreme beach inundation values. That is why finally the 9th formula with correction for statistical error to predict beach inundation was selected:

$$BI_{MS} = \frac{R_2}{h} \cdot 100\% + SYE \tag{7}$$

$$R_2 = 0.53\beta\sqrt{(H_0 L_0)} + 0.58\xi_0\sqrt{\frac{H_0^3}{L_0}} + 0.45 + S \tag{8}$$

Because natural processes around the beaches may be modified by various constructions, it could be difficult to predict correctly beach inundation without adding any additional component, in some percentage interpreting human influence in the analyzed areas, to the equations. However, in some situations wave run-up formulas are a good alternative to very complicated three-dimensional models and are successfully used by coastal engineering brands.

References

Aniśkiewicz P, Benedyczak R, Furmańczyk K, Andrzejewski P (2016) Validation of empirical wave run-up formulas to the Polish Baltic Sea coast. J Coastal Res 75(sp1):243–247

Battjes JA (1974) Surf similarity. Coast Eng Proc 1(14):466–480

Bugajny N, Furmańczyk K (2014) Dune coast changes caused by weak storm events in Miedzywodzie, Poland. J Coastal Res 70(sp1):211–216

Bugajny N, Furmańczyk K, Dudzińska-Nowak J, Paplińska-Swerpel B (2013) Modelling morphological changes of beach and dune induced by storm on the Southern Baltic coast using XBeach (case study: Dziwnow Spit). J Coastal Res 65(sp1):672–677

Bugajny N, Furmańczyk K, Dudzińska-Nowak J (2015) Application of XBeach to model a storm response on a sandy spit at the southern Baltic. Oceanol Hydrobiol Stud 44(4):552–562

Cabanes C, Cazenave A, Provost C (2001) Sea level rise during past 40 years determined from satellite and in situ observations. Science 294(5543):840–842

Church JA, White NJ (2006) A 20th century acceleration in global sea-level rise. Geophys Res Lett 33(1)

Church JA, White NJ (2011) Sea-level rise from the late 19th to the early 21st century. Surv Geophys 32(4–5):585–602

Cieślikiewicz W, Paplińska-Swerpel B (2008) A 44-year hindcast of wind wave fields over the Baltic Sea. Coast Eng 55(11):894–905

Díaz-Sánchez R, López-Gutiérrez JS, Lechuga A, Negro V (2014) Runup variability due to time dependence and stochasticity in the beach profiles: two extreme cases of the Spanish coast. J Coastal Res 70:1–6

Douglas BC (1991) Global sea level rise. J Geophys Res, Oceans 96(C4):6981–6992

Galvin CJ (1972) Wave breaking in shallow water. In: Waves on beaches and resulting sediment transport, pp 413–456

Guza RT, Thornton EB (1984) Swash oscillations on a natural beach. J Geogr Res Oceans 87 (NC1):483–491

Hedges TS, Mase H (2004) Modified Hunt's equation incorporating wave setup. J Waterw Port Coast Ocean Eng 130(3):109–113

Holman RA, Sallenger AH Jr (1985) Setup and swash on a natural beach. J Geophys Res 90 (C1):945–953

IPCC (2007) Climate change 2007. The physical science basis. In: Solomon S, Qin D, Manning M, Chen Z, Marquis M, Averyt K, Tignor M, Miller HL (eds) Contribution of working group I to the fourth assessment report of the intergovernmental panel on climate change. Cambridge University Press, New York, p 996

Iribarren CR, Nogales C (1949) Protection des ports. In: Proceedings XVIIth International Navigation Congress, Section II, Communication 4, Lisbon, pp 31–80

Jones PD, Moberg A (2003) Hemispheric and large-scale surface air temperature variations: an extensive revision and an update to 2001. J Clim 16(2):206–223

Łabuz TA (2013) Sposoby ochrony brzegów morskich i ich wpływ na środowisko przyrodnicze polskiego wybrzeża Bałtyku. WWF report, pp 24–109

Leppäranta M (1981) An ice drift model for the Baltic Sea. Tellus 33(6):583–593

Leppäranta M, Myrberg K (2009) Physical oceanography of the Baltic Sea. Springer Science & Business Media

Mase H (1989) Random wave runup height on gentle slope. J Waterw Port Coast Ocean Eng 115 (5):649–661

McGranahan G, Balk D, Anderson B (2007) The rising tide: assessing the risks of climate change and human settlements in low elevation coastal zones. Environ Urbanization 19(1):17–37

Nicholls RJ, Cazenave A (2010) Sea-level rise and its impact on coastal zones. Science 328 (5985):1517–1520

Nicholls RJ, Wong PP, Burkett V, Codignotto J, Hay J, McLean R, Ragoonaden S, Woodroffe CD, Abuodha PA, Arblaster J, Brown B (2007) Coastal systems and low-lying areas

Nielsen P, Hanslow DJ (1991) Wave runup distributions on natural beaches. J Coastal Res 7 (4):1139–1152

Ortega-Sánchez M, Fachin S, Sancho F, Losada MA (2008) Relation between beachface morphology and wave climate at Trafalgar beach (Cádiz, Spain). Geomorphology 99(1): 171–185

Paprotny D, Andrzejewski P, Terefenko P, Furmańczyk K (2014) Application of empirical wave run-up formulas to the Polish Baltic Sea coast. PLoS ONE 9(8):1–8

Rotnicki K (1995) The coastal zone—present, past and future. J Coastal Res 22:3–11 (Polish Coast, Past Present and Future. Quaternary Research Institute, Adam Mickiewicz University, Poznań)

Ruggiero P, Komar PD, McDougal WG, Beach RA (1996) Extreme water levels, wave runup and coastal erosion. Coast Eng Proc 1(25):2793–2805

Staniszewska M, Boniecka H, Gajda A (2014) Prace pogłębiarskie w polskiej strefie przybrzeżnej–aktualne problemy. Inżynieria Ekologiczna

Stockdon HF, Holman RA, Howd PA, Sallenger AH (2006) Empirical parameterization of setup, swash, and runup. Coast Eng 53(7):573–588

van der Meer JW (1998) Wave run-up and overtopping. In: Pilarczyk KW (ed) Dikes and revetments: design, maintenance and safety assessment. AA Balkema, Rotterdam, pp 145–159

Vousdoukas MI, Wziatek D, Almeida LP (2012) Coastal vulnerability assessment based on video wave run-up observations at a mesotidal, steep-sloped beach. Ocean Dyn 62(1):123–137

Williamson P (1992) Reducing uncertainties. Coastal connections. IGBP, Stockholm, pp 19–21

Wróblewski A (1994) Analysis and long-term forecast of sea-levels along the Polish Baltic Sea coast. Part II. Annual mean sea-levels—forecast to the year 2100. Oceanologia 36:107–120

Zeidler B, Wróblewski A, Miętus M, Dziadziuszko Z, Cyberski J (1995) Wind, wave, and storm surge regime at the Polish Baltic coast. J Coast Res (SI 22):3–11

The Impact of the Sopot Pier Marina on the Local Surf Zone

Anna Przyborska, Jaromir Jakacki and Szymon Kosecki

Abstract An accelerated accumulation of sand in the area of the Sopot Pier Marina has been observed since 2010. The beach near the pier has grown by several meters and local depths have been decreasing. Before construction of the marina, spit type shoreline form had sometimes formed. But it has never begun to grow in the way it has grown after 2010, as a result of disturbed sediment transport in the local surf zone. The main goal of this work is to estimate where this process is going on. To assess the influence of the marina on the longshore drift, geometrical analysis of the existing marina has been performed and numerical model of sediment transport has been implemented. A simple analysis of the Sopot Pier dimensions suggests that the final form of growing spit will be tombolo. For the purpose of having an additional point of view, a numerical model of sediment transport based on MIKE by DHI product has been developed. The model results show that the Sopot Pier Marina has a strong influence on the sediment transport by generating a wave shadow zone. Consequently, it disturbs the continuity of sediment transport along the beach, which as a result impacts the costal line. The model indicates that 80% of this sand was transported from the area located southeast of the Sopot Pier, whereas the remaining 20% from the northwest area. Simulations showed that accumulated sand came from local surf zone and beaches. This result has been confirmed by bathymetry measurements performed by the Maritime Office in Gdynia.

Keywords Sopot pier · Tombolo · Sediment transport · Local coastal zone
Shoreline

A. Przyborska (✉) · J. Jakacki · S. Kosecki
Institute of Oceanology, Polish Academy of Sciences, Sopot, Poland
e-mail: aniast@iopan.gda.pl

J. Jakacki
e-mail: jjakacki@iopan.gda.pl

S. Kosecki
e-mail: skosecki@iopan.gda.pl

T. Zielinski et al. (eds.), *Interdisciplinary Approaches for Sustainable Development Goals*, GeoPlanet: Earth and Planetary Sciences,
https://doi.org/10.1007/978-3-319-71788-3_8

1 Introduction

The Sopot Pier, one of the flagships of the Polish coast, is located at the end of the Monte Casino Street in Sopot, Poland. With a length of over 500 m it is the longest wooden pier on the coast of the Baltic Sea, and it is also one of the biggest tourist attractions of Tricity (metropolitan area consisting of three cities: Gdańsk, Sopot and Gdynia). Its history dates back to 1827 when the first wooden structure was built there to become a popular venue for the customers of the bathing facility developed by Jean Haffner—a medical doctor and the founder of the first spa located in Sopot. In the following years, the pier was extended and the largest reconstruction took place in 1927. In the 1990s, a reinforced concrete head was added. In 2005 a place for vessels was established and in 2011 a yacht marina construction was completed at the end of the pier. Now the Sopot Pier is one of the most popular tourist attractions; in summer, tourists and athletes take part in numerous tournaments and various sports competitions. The venue attracts between 1.5 and 2 million people each year.

In recent years the beach at the Sopot Pier has been growing by several meters per year (Fig. 1). Recently, the shoreline has reached the first lower deck and in summer stagnant water appears and algal blooms emit odour (Kiwnik 2014), which generates considerable costs for the city. These changes are troublesome, as sometimes it is necessary to close the beach due to biological contamination.

2 Simple Dimensions Analysis

It is natural that in a coastal zone the bottom material is transported during wave braking. Wave energy is converted into orbital motion of water and bottom material is taken out (of course if possible, if there is enough energy for it) and moved forward and back. This process has been shown in Fig. 2 as green arrows.

Fig. 1 Changes in the shoreline near the Sopot Pier (left side: January 2008, right side: November 2016) Google earth V 7.1.8.3036

This motion provides an average sediment transport along the shoreline (represented as yellow arrows). That local sediment transport heavily depends on energy of waves and their direction. The marina structure, projecting from the shoreline, has acted as a powerful, detached breakwater. When the waves reaching the coastal zone are damped (and refracted; however, the influence of refraction decreases with increasing wave energy) by construction of the marina and a salient area is created between the marina (marked in Fig. 2 as 'disturbed sediment transport area'). As a result this perturbation salient sand feature (also known as spit) began to grow.

Types of shoreline forms which could appear under the influence of marina have been described in the scientific literature (Pruszak and Skaja 2014; Chasten et al. 1993) and they mostly depend on the alongshore length of the marina (L) and the distance between the marina wall and the shoreline (X) (Fig. 3). Whether the formed shape is of spit (salient type) or of tombolo type depends on the ratio of parameters (L/X). The experimental laboratory studies (Suh and Dalrymple 1987) show that for $L/X > 1.3$ there is a high probability of tombolo, whereas for $0.5 < L/X < 1.3$ it is highly likely that shoreline will take a salient form. The rate

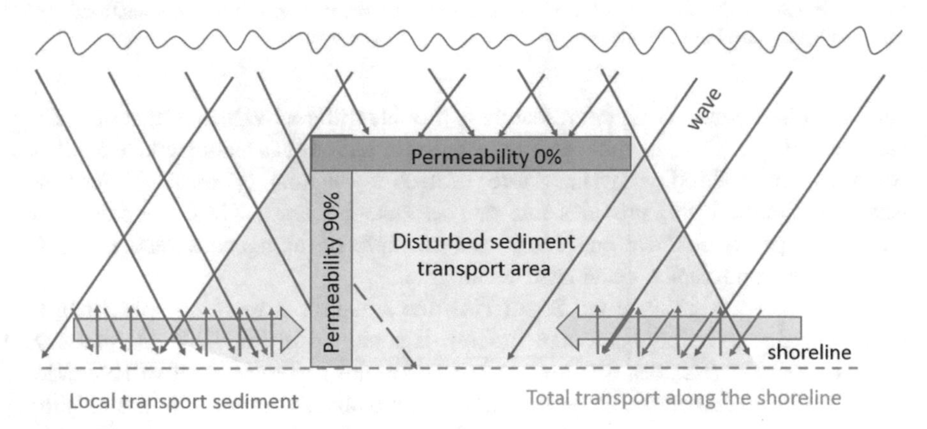

Fig. 2 Schematic representation of the impact of the new marina in Sopot on local sand transport along shoreline

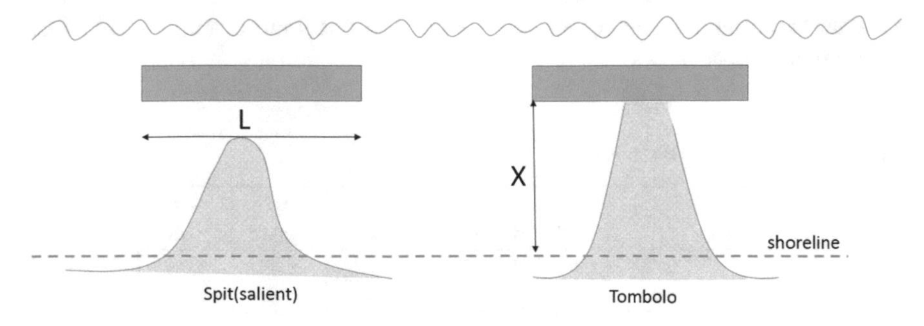

Fig. 3 Shoreline forms: spit and tombolo

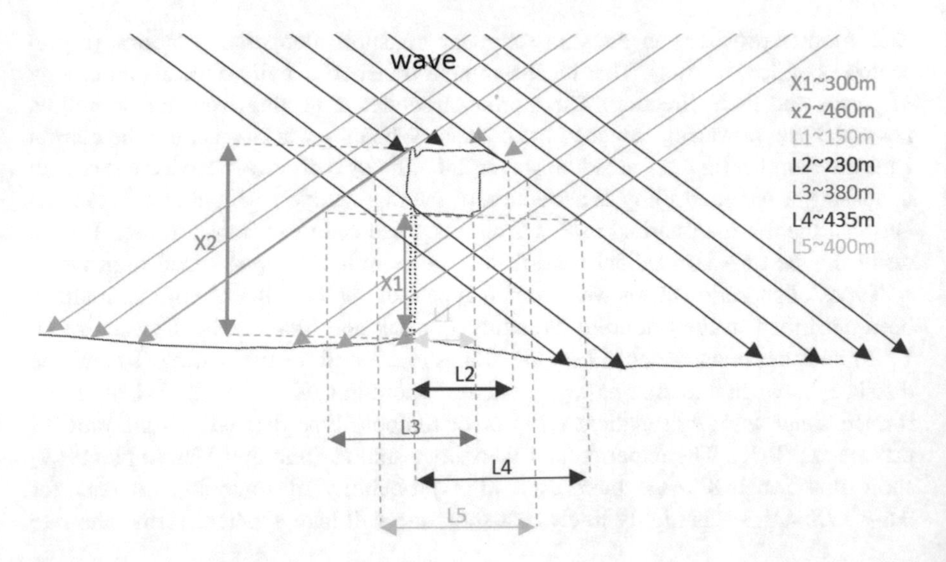

Fig. 4 Dimensions, which could have influence on spit/tombolo formation in the analyzed area (the Sopot Pier and the marina)

L/X is not hardwired. Different researchers provide different values of this rate. The smallest value of this rate that causes tombolo formation was estimated for shallow water as 0.67 (Gourlay 1981). Shore Protection Manual (Coastal Engineering Research Centre 1984) provides this rate as greater than 1. Thus, we can say it strongly depends on local conditions, for example on alongshore sand transport rate, bottom topography, dominated winds, etc.

When we try to analyze the Sopot Pier, the situation is more complex than in laboratory due to its complex construction. It is shown in Fig. 4. When real L/X ratio is about 0.5 (L2/X2), it is important to take into account 'effective L/X ratio' which could be calculated based on surface wave shadow created by the existing marina. Thus, the L/X ratio depends on the wave direction and it is in the range between 0.33 and 1.45 (Table 1). The greatest value of L/X is 1.45 and it is nearly 10 percent greater than the value presented before (Suh and Dalrymple 1987). It leads to a conclusion that construction of the marina disturbed local sediment transport and tombolo is being formed. To confirm this finding, a numerical model has been developed.

Table 1 Ratios L/X shown in Fig. 4

		L1	L2	L3	L4	L5
Ratio L/X		150	230	380	435	400
X1	300	0.50	0.77	1.27	1.45	1.33
X2	460	0.33	0.50	0.83	0.95	0.87

3 Numerical Model

For better understanding of this process, a numerical model of sediment transport has been adapted for this area. The model is based on MIKE by DHI tools. It is a commercial product developed by DHI (Danish Hydraulic Institute) for assessment of non-cohesive sand transport rates, related to initial bed level changes and morphological evolution. The model was implemented as a coupled system of the following modules (MIKE by DHI docs. 2010):

- Hydrodynamic Module (HD),
- Spectral Wave Module (SW),
- Sand Transport Module (ST).

The Hydrodynamic Module is the main part of the implemented system. It solves Reynolds-averaged Navier-Stokes (MIKE by DHI docs. 2014) equations (RANS) for an incompressible medium with the Boussinesq assumption and shallow water approximation and it is based on a finite volume method on an unstructured mesh. The main module was used to calculate surface elevation and horizontal and vertical currents structure, which in the given applications were wave-driven, and the sediment transport module was finally used to estimate the sediment transport field.

The Spectral Wave Module solves the wave action balance equation using a directional decoupled parametric formulation (Holthuijsen et al. 1989). The wave transformation incorporates the effects of energy dissipation due to wave breaking and bottom friction, energy transfer due to depth refraction and diffraction and energy supply due to action of wind (MIKE by DHI docs. 2013). The depth-induced wave breaking criteria, which are used in MIKE, are based on classical dependence between height of breaking wave (H) and depth of water (h) in the breaking point:

$$H = \gamma h, \tag{1}$$

where the parameter $\gamma = 0.8$ represents a wave height to depth ratio, this value is suitable for most wave breaking cases (Battjes and Janssen 1978). The wave-induced currents were generated by the gradient of radiation stress present in the surf zone.

The Sand Transport Module calculates sand transport rate on a flexible mesh covering the area of interest on the basis of the hydrodynamic data obtained from simulation with the Hydrodynamic Module and potentially available wave data (provided by MIKE SW) together with the information about characteristic of the bed material.

3.1 Model Setup and Numerical Parameters

Mike by DHI is a well-known and very well documented (MIKE by DHI docs. 2014) product; therefore, we are going to describe only all the important features used in our implementation, while other configuration parameters and settings are going to be presented in the table.

The model was placed on a grid of variable resolution that allows increasing the resolution in the vicinity of the pier (Fig. 6). The unstructured grid enabled a variable horizontal resolution. The model grid consists of 4847 elements and 2615 nodes. The smallest cell in our domain had a horizontal resolution of ca 2 m and the largest cell had a dimension of ca 200 m. Smaller cells covered parts close to the pier. The model bathymetry has been generated from experimental data provided by the Polish Maritime Office. The bathymetry, model grid and domain are shown in Fig. 5.

To have the model working properly it is important to provide lateral boundary conditions. The speed and direction of sea currents were derived from a HIROMB-High Resolution Model for the Baltic Sea (Funkquist and Ljungemyr 1997); temperature and salinity were taken from the same source (although variability of temperature and salinity have negligible influence on sand transport,

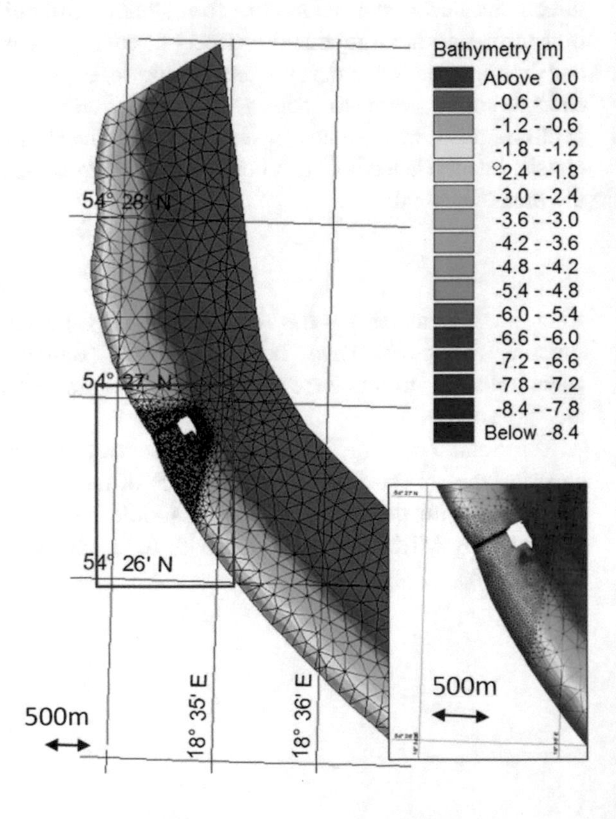

Fig. 5 The model domain and bathymetry with variable resolution grid

Table 2 Parameters used in the model

Parameter	Model option/value
Horizontal resolution	Mesh grid in Fig. 6 max size of cell 200 m
Simulation periods	04.2010–05.2011
Maximum time step	120 s
Bathymetry source	Maritime Office Gdynia
Hydrodynamic module	
Flood and dry	Included
Density	Salinity and temperature dependent
Horizontal eddy formulation	Constant eddy formulation
Bed resistance	Manning number
Coriolis forcing	Included
Wind forcing	Based on HIRLAM
Temperature and salinity of water, current speed and direction	HIROMB
Wave radiation	From SW simulation
Critical CFL value (Courant–Friedrichs–Lewy)	0.8
Initial conditions—cold start	
Surface level	0 m
Velocities	0 m/s
Sand transport module	
Model type	Wave and current—threshold thickness 0.0005 m
Sediment properties	Porosity 0.4 Grain diameter 0.25 mm Grading coefficient 1.1
Waves	Wave field from SW simulation
Morphology	Max bed level change 0.5 m/day Includes feedback on hydrodynamics, wave and sand transport calculation Bank erosion type—slope failure, 30 deg
Initial condition	Layer thickness constant value 1 m
Boundary condition	Zero sediment flux gradient for outflow, zero bad change for inflow
Spectral wave module	
Spectral formulation	Directional decoupled parametric formulation
Spectral sector discretization	Number of directions 9 Minimum directions 315 deg Maximum direction 135 deg
Water level variation	From HD simulation
Current variation	From HD simulation

(continued)

Table 2 (continued)

Parameter	Model option/value
Diffraction	Included
Wave breaking	Type specified gamma constant value 0.8
Bottom friction	Sand grain size, d 50 = 0.00025 m
Initial condition	Zero spectra
Boundary condition	Wave parameters: Significant wave height Peak wave period Mean wave direction

we applied them because the hydrodynamic module requested it at the boundaries). Forces for the Spectral Wave Model include: water level and current variation from HD simulation, significant wave height, peak wave period, mean wave direction. All the wave data were obtained from WAM operated by the Maritime Institute in Gdańsk. The WAM model was driven using HIRLAM (High Resolution Limited Area Model) atmospheric data and the same data was implemented as top boundaries for hydrodynamic model. Time resolution of the boundaries data was 1 h, horizontal resolution of WAM data was 5 km, HIROMB provided all the data with 1 nm (nautical mile, which is equal to about 1.8×1.8 km) grid, horizontal resolution of the HIRLAM data was about 0.15° (Sass et al. 2002). In the adapted model, all the data have been interpolated into model grid.

As was mentioned before, other parameters and settings are included in Table 2.

4 Local Conditions

Dominant wind directions in the pier area range from northwest to east. Figure 6 shows the wind rose for this area (winds from HIRLAM model). The chart shows the frequency and speed of wind blowing from each direction. It is not a typical situation that prevails in the southern Baltic Sea (there are winds ranging mostly from west to north). The form of the wind rose results from the shape of the shoreline in this region. Winds from the west and northwest are suppressed by a high planetary boundary layer formed by significant transition zone ranging of several kilometers from southeast to northwest. That zone, which is an area of increased roughness, resulting in an increase in turbulences, causes dissipation of wind energy and consequently damped winds from these directions.

The presence of winds in the implemented model does not have a significant impact on the bottom sand transport in the coastal zone because it is caused by wind-driven waves but generated in a different spatial area (mostly in the open sea). The main physical quantity responsible for the sand transport is energy (which is proportional to the square of the wave amplitude) dissipated from braking waves. As mentioned above, the energy carried out by high waves is the most effective.

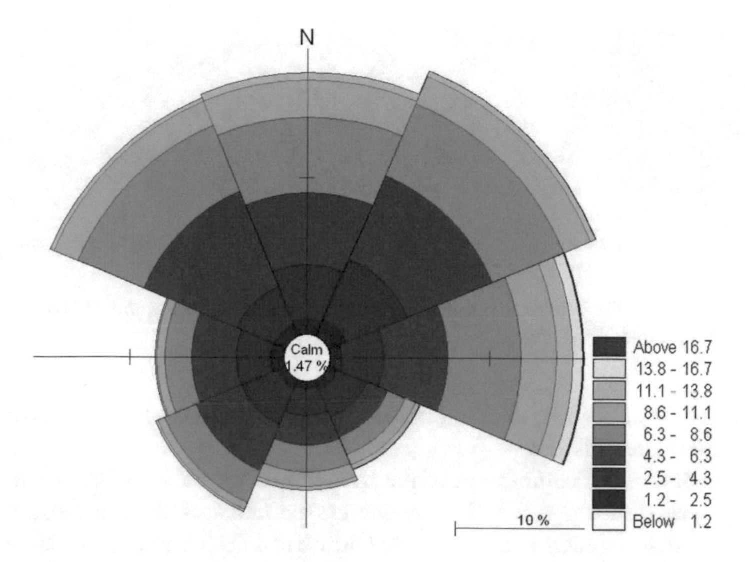

Fig. 6 Typical distribution of wind speed (m/s) and direction in the area of the pier

Such waves can be formed only in the open sea and due to the bathymetry of the Gulf of Gdansk they can reach the pier.

The Sopot shoreline is partially sheltered from the open sea waves by the Hel Peninsula. There are important directions of waves—from the North to the East—that have strong influence on sediment transport in the Sopot Pier area. The significant waves (that have enough energy) are formed at such a great distance from the shore in the deeper area of the sea. Waves from the west are diffracted on the Hel Peninsula and move to the area of the pier.

The currents in this area are mainly controlled by sea level difference between the Gulf of Gdansk and the Puck Bay. In spite of the fact that western and northern winds prevail there, north-west circulation occurs in the vicinity of the pier. Maximum of these currents may be several cm/s and they have a negligible effect on the sediments transport in the surf zone near the pier.

5 Deposition of Bottom Material in the Sopot Pier Area

Numerical integration has been performed only from April 2010 to May 2011 because of limited number of available experimental input data for the model. The validation process of sediment transport is also difficult because of restricted and inconsistent data. The Maritime Office in Gdynia has an obligation of performing bathymetry surveys in the Sopot Pier area, but the gathered data are of different resolution and cover different areas. Figure 7 shows a different spatial resolution of measured bottom topography. Although years 2010 and 2012 are pretty well

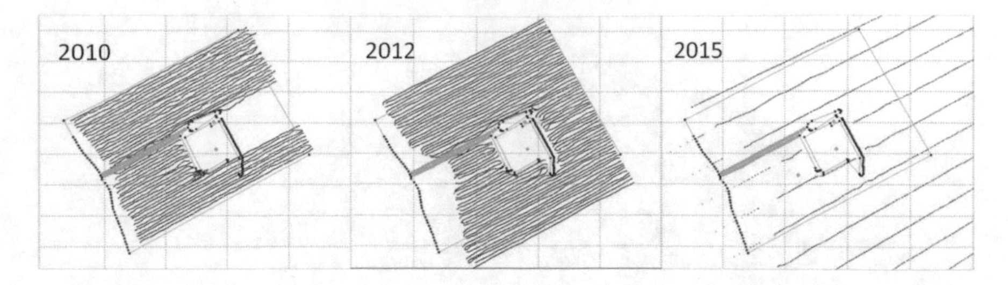

Fig. 7 Location of the bathymetry measurement points taken into account when determining seabed accretion rate in the area of the pier

documented, year 2015 has limited profiles of measurements. It is also visible, that covered areas are different.

For the purpose of comparison of the model results with experimental data only the above years, shown in Fig. 7, have been taken into account. For those years we have selected a common area, which is shown in Fig. 8 and the available in situ data permitted to compare the accumulation of sediments in the selected area.

Based on the bathymetric data from the Maritime Office for the covered area, the deposited sand volume in the selected area has been estimated and the year 2010 has been taken as reference year. Based on that, the trend of growing of sediment volume was estimated and plotted (Fig. 9). Numerical model integration provided increased sediment volume in the vicinity of the Sopot Pier area, but for much shorter time and with much better temporal resolution. This result is also plotted in Fig. 9 as the model result (the green points). One-year simulation provided similar

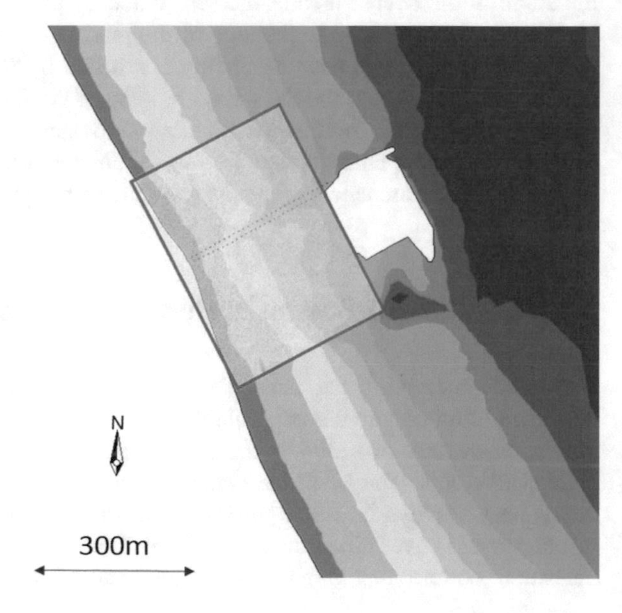

Fig. 8 The area for which sediment deposition rate was specified

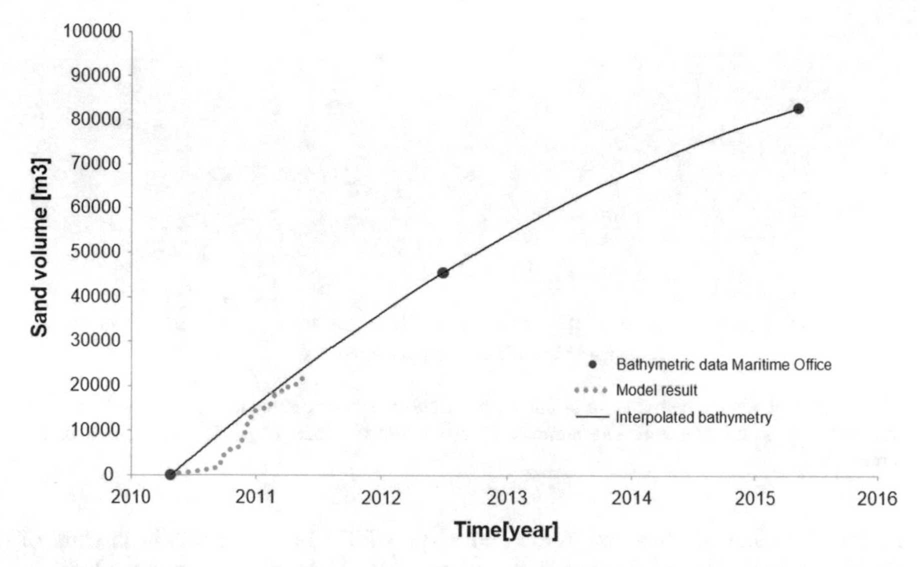

Fig. 9 Comparison of sediment deposition rate obtained from the numerical model and determined on the basis of data from the Maritime Office

collected volume of sediment in the end of simulation (clearly visible in Fig. 9 in the end of green points—between years 2011 and 2012).

To validate the bottom material accumulation, we should compare the measured and the modeled curves in Fig. 9, but at this stage we are able to compare only one point. Thus, it is impossible to find valuable number that will represent the comparison properly.

In the, years 2010–2015, there was a nonlinear increasing trend in the volume of sediment (based on the Maritime Office data). In 2015, the volume of sediment deposited in the area selected for the analysis in relation to the reference year was already at a level of 82,000 m^3 (Fig. 9), giving an average deposition rate of about 16,000 m^3/year. It is important to add that we are aware that the trend is not constant, however, this value could be used for estimation of potential volume of sand needed to be removed to restore the bathymetry from previous years. We also try to avoid making forecast which would be more of guessing than prediction. The sediment deposition curve tends to saturation. The nature of the curve shows that the process will continue to decelerate until it reaches equilibrium—a fully developed tombolo.

Numerical simulation was carried out for the period between April 2010 and May 2011 and if we will take a look at this part of Fig. 9, in the beginning of simulation the sediment deposition rate was low. That was due to lack of sufficient wave energy necessary for the meaningful transport of bottom material. It was a time of weak winds in late spring and summer. An increase in accumulated

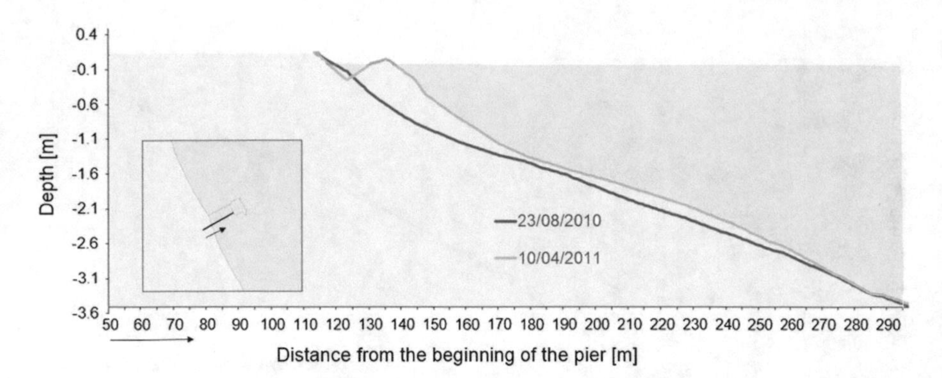

Fig. 10 Change in the bathymetric profile perpendicular to the shoreline marked by the black line, the blue line is the profile in late summer of 2010, and the orange profile—at the end of the simulation

sediment volume is observed in the end of year 2011 and later and in the end of simulation the volume became more comparable to the data from the Maritime Office. Also, when we take a look at the bathymetric profile perpendicular to the shoreline (Fig. 10), it shows that sediment transport is the most intensive in the shore zone and reduces with distance from the shoreline.

The greatest change in this bathymetry profile is nearly 1 m. Based on the presented model results we can say that the model reflects the main important changes in the bottom topography in the Sopot Pier area.

For confirmation of explanation of an argument, that the most energetic waves have the main influence on accumulation of the bottom material in the analyzed area, we have made a special experiment. The first step is the characterization of the forcing. We have defined 'significant events' as the event exceeding a minimum significant wave height ($Hs > 0.2$ m) and wave direction from north-west to south-east. This assumption is based on the definition of wave energy, which is proportional to the square of amplitude:

$$E \approx \int_{t_1}^{t_2} H_{rms}^2 \mathrm{d}t \tag{2}$$

where $(t_1 - t_2)$ is the time of appearing of "significant events" and H_{rms} represents root mean square wave height. The threshold of significant events defined the 33% of the data set and more than 98% of the energy. For this data set we prepared "short simulation". The simulation results for a full time period (blue dots) well overlap with results of the simulation carried out only for "significant events" (red dots in Fig. 11). The final results in both cases are the same.

The aerial view of the modeled area in case of different incoming wave direction (Fig. 12) shows that the marina generates a wave shadow zone (an area where

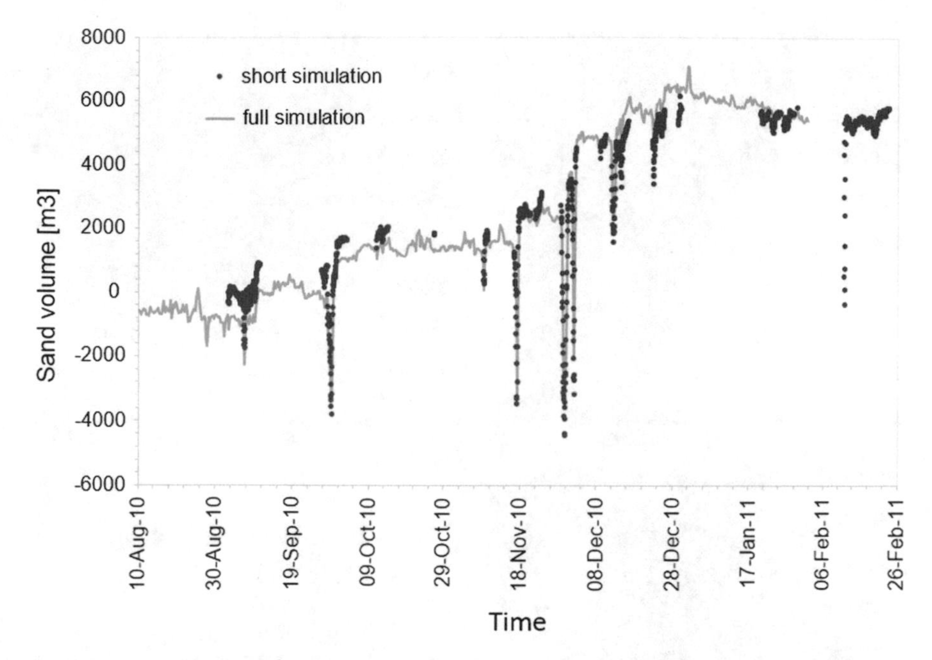

Fig. 11 Changes in the volume of sand in the selected area (Fig. 8) for two different simulations (blue dots: simulation results for a full time period, red dots: short simulation only for "significant events")

waves do not reach the shore but are suppressed by the marina). Reduction in wave height and deflection of the waves behind the breakwater have dramatic impact on the wave-driven currents and the littoral drift behind the breakwater. This drift is reduced in the center of the shadow zone (behind the breakwater), while it is increased near the boundaries of the shadow zone. As a result, the marina causes sediment accumulation in the shadow zone, while erosion occurs in the neighboring coastal profiles. This shadow disturbs the continuity of sediment transport along the beach, the consequence of which is formation of spit and then tombolo.

The developed model could also provide the area of the marina impact on the shore, which is shown in Fig. 13 as two red straight lines.

According to Kapiński et al. (2004) the sediment transport along the shore near the Sopot pier was estimated at ca. 20,000 m^3/year and it is directed southward.

In our sediment transport model the volume of carried sediment is comparable, but the direction of transport is different (Fig. 14). Our simulations show that 80% of the sand accumulated near the pier come from local beaches located southeast of the pier. The remaining 20% was transported from the north-west side of the pier (note: the wind direction is the direction from which it originates (Fig. 6) but current direction shows the direction of flow and this nomenclature is used for direction of waves and longshore drift). The direction and volume of sand transport (on the left) and significant wave height and its direction (on the right side) are shown in Fig. 15. The figures show the frequency and volume of sediment transport

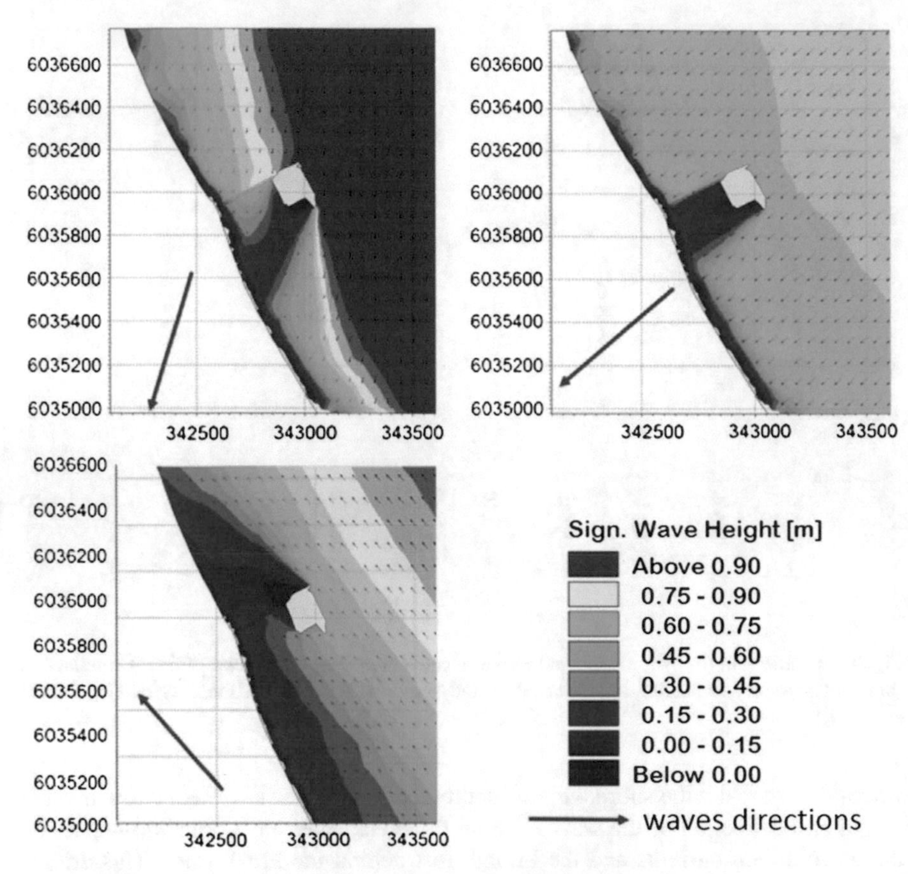

Fig. 12 Masking/shadow waves near the marina for different waves directions

and the frequency of waves to each direction. The direction of sediment transport corresponds to directions of the local most energetic waves quite well.

The sediments transport directions do not reflect only years the 2010–2011. We have also analyzed the data provided by the Maritime Office. To confirm the assessed sand transport in the Sopot Pier area we distinguished two parts (denoted as GDA and GD in Fig. 16) from the common area shown in Fig. 8. Based on the measurements data, for years 2010, 2012 and 2015 the sediment accumulation in the GA and GDA regions on both sides of the spit was calculated. The areas are symmetrical to the maximum of the spit and their fields are equal (image and area chart in Fig. 16). Both of them get shallower and accumulate bottom material in years 2010–2015. Figure 16 shows that more sediment is accumulated in the GDA area in the period 2010–2012 and 2012–2015 (marked red) than in the GA area (marked green).

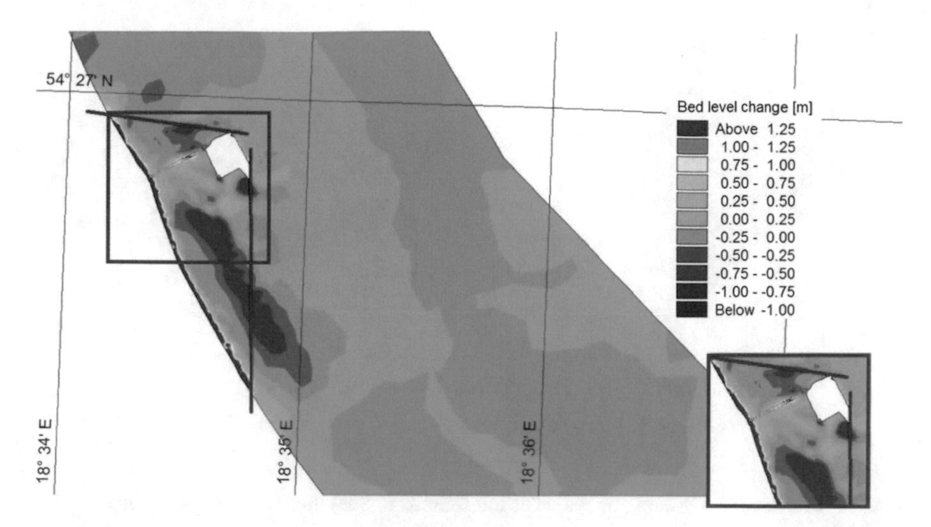

Fig. 13 Changes in bed elevation/bathymetry and the marina impact area (marked with red lines)

Fig. 14 Sediment transport near the Sopot pier (Google earth V 7.1.8.3036. [November 4, 2016])

However, this accretion of sediment in the GDA area compared to sediment growth in the GA area decreases over time. It leads to a conclusion that deposition of sand in the GDA area will reach equilibrium earlier.

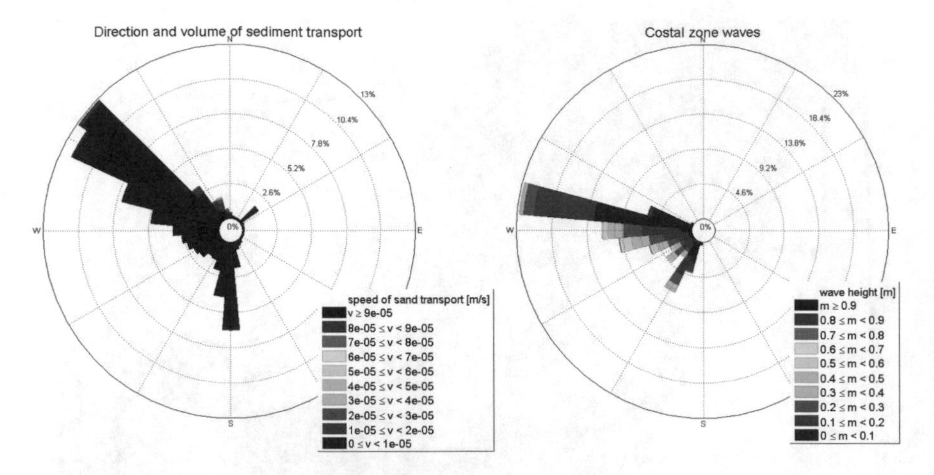

Fig. 15 Direction and volume of sediment transport (left side) and wave height (right)

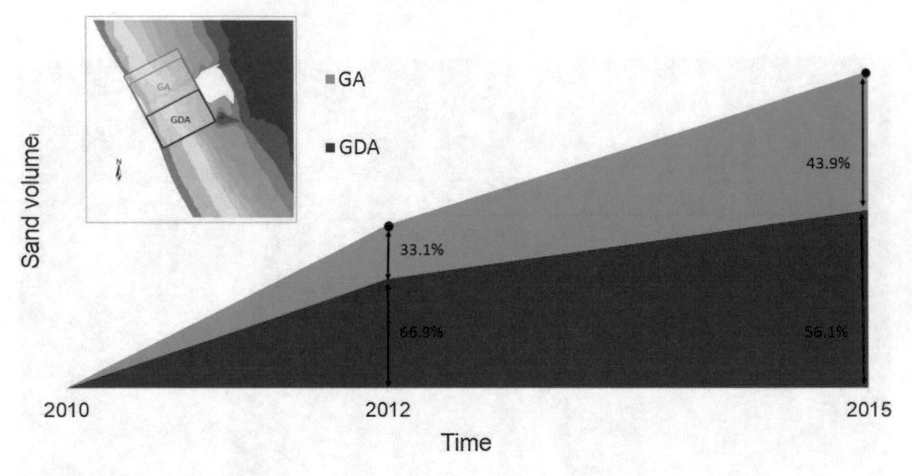

Fig. 16 Sediment accumulation in the period 2010–2015 determined on the basis of data from the Maritime Office for two selected areas—GA (red color) and GDA (green color)

6 Conclusions

The Sopot marina generates a wave shadow zone. This shadow disturbs the continuity of sediment transport along the beach, the consequence of which will be a fully developed tombolo. The implemented numerical sediment transport model for the analyzed area reflects well the sediment transport. The results show that the volume of accumulated sediment in the area selected for analysis was already at the level of 82,000 m^3 for the period 2010–2015. But this value depends heavily on the region. The sediment deposition curve tends to saturation and the process will decelerate until it reaches equilibrium—a tombolo. The analysis also shows that most changes occur in the sediment accumulation process in autumn and winter, when strong winds generate significant waves with sufficient energy for alongshore sediment transport. The strong winds (usually) do not appear in summer. Simulations also showed that about 80% of the sand accumulated near the pier has come from local shore zone located southeast of the pier and 20% was transported from the north-west side of the pier. The direction of sediment transport corresponds to directions of local waves and it was confirmed by analysis of bathymetries provided by the Maritime Office.

References

Battjes JA, Janssen JPFM (1978) Energy loss and set up due to breaking of random waves

Chasten MA, Rosati JD, McCormic JW, Randall RE (1993) Engineering design guidance for detached breakwaters as shoreline stabilisation. Vicksburg, Mississippi

Coastal Engineering Research Centre (1984) Shore protection manual. Washington

Funkquist L, Ljungemyr P (1997) Validation of HIROMB during 1995–1996. SMHI Oceanografi Nr. 67

Gourlay MR (1981) Beach processes in the vicinity of offshore breakwaters. Fifth Australian conference on coastal and ocean engineering

Holthuijsen LH, Booij N, Herbers THC (1989) A prediction model for stationary, short-crested waves in shallow water with ambient currents. Costal Engr

Kapiński J, Ostrowski R, Skaja M, Szmytkiewicz M (2004) Wpływ planowanej przystani jachtowej w Sopocie na brzeg morski (in Polish). Inżynieria Morska i Geotechnika, nr 4:180–185

Kiwnik J (2014) Powiększa się łacha piasku przy molo. To przez betonową marinę (in Polish). http://trojmiasto.wyborcza.pl/. Gazeta Wyborcza Trójmiasto

MIKE by DHI docs. (2010) Scientific Documentation, MIKE 21 & MIKE 3 FLOW MODEL FM Hydrodynamic and Transport Module. Danish Hydraulic Institute

MIKE by DHI docs. (2013) MIKE 21 SW—spectral waves FM short description. Danish Hydraulic Institute

MIKE by DHI docs. (2014) Scientific documentation MIKE 21 & MIKE 3 FLOW MODEL FM hydrodynamic and transport module scientific documentation. Danish Hydraulic Institute

Pruszak Z, Skaja M (2014) Problemy dynamiki i ochrony brzegu morskiego (in Polish). IBW PAN

Sass BH, Nielsen NW, Jørgensen JU, Amstrup B, Kmit M, Mogensen KS (2002) The operational DMI-HIRLAM system 2002-version. Technical, Danish Meteorological Institute, Copenhagen

Suh K, Dalrymple R (1987) Offshore breakwaters in laboratory and field. J Waterw Port Coast Ocean Eng

Arsenic in the Environment of the Baltic Sea—A Review

Marta Szubska

Abstract Arsenic is an ubiquitous element, naturally occurring in all compartments of the environment. In marine ecosystems its concentrations are even one order of magnitude higher than in terrestrial environments, as arsenic entering the global cycle is eventually transported to the oceans where it bounds with sediments. It is toxic to many marine organisms, and consequently via the food web it might create risk for fish consumers. Seafood is the main source of arsenic in human organism; therefore, investigating arsenic transformations in marine environment is very important. Baltic Sea is a very specific water body. Geographical position, geological development, hydrographical features and physical drivers together make it very susceptible to pollutants, including arsenic contamination. In the case of Baltic Sea there is an additional source of arsenic—arsenic containing chemical weapon deposited on the sea bottom after the World War II. The aim of this article was to summarize the knowledge on arsenic behavior in the Baltic Sea gained by many researchers through the years. An effort was given to synthesize the information regarding arsenic concentrations in Baltic water, bottom sediments and marine organisms. Comparing with other marine areas, arsenic occurs in the Baltic Sea in relatively low concentrations, however in some regions (Bothnian Bay) the contamination with arsenic is significant and can pose a threat to organisms.

Keywords Arsenic · Baltic Sea · Marine environment · Marine sediments

1 Introduction

1.1 Baltic Sea

Baltic Sea is one of the biggest brackish water bodies on earth. It is a non-tidal, epicontinental sea—situated on a continent, not between continents, what makes it

M. Szubska (✉)
Institute of Oceanology of the Polish Academy of Sciences, Sopot, Poland
e-mail: szubi@iopan.gda.pl

T. Zielinski et al. (eds.), *Interdisciplinary Approaches for Sustainable Development Goals*, GeoPlanet: Earth and Planetary Sciences,
https://doi.org/10.1007/978-3-319-71788-3_9

partially cut off from the oceans (Preis 1997). Its shape and location, spreading between subarctic to moderate climate longitudinally and oceanic to continental latitudally, make it unique on a global scale. Geographical position, geological development, hydrographical features, and physical drivers together create the Baltic Sea environment. The presence of shallow basins between particular basins plays role in transport of water masses, chemical substances and sedimentary material. Therefore, different areas of the Baltic Sea are characterized by diversified bottom types and sediment composition. Also salinity decreases in the north-eastern direction—the highest salinity is observed in Kattegat (18–28), slightly lower in Danish Straits (10–20), moderate in the Baltic Proper (6–10) and the lowest (4) in the Gulf of Finland and Bothnian Bay (Snoeijs-Leijonmalm et al. 2017; Uścinowicz 2011).

Baltic Sea is directly surrounded by nine countries, with additional five laying within the drainage area, and each of them has its impact on the Baltic Sea environment (Fig. 1). Nearly 1,750,000 km^2 drainage area is about 4 times bigger than the area of Baltic Sea itself, heavily industrialized and relatively polluted (Uścinowicz 2011; Preis 1997). Large input of freshwater results in low salinity and characteristic salinity gradient due to mixing with saline waters entering via narrow and hollow straits of the Kattegat. This limited inflow of oceanic water masses results in quite long residence (turnover) time, estimated on 30–40 years for the whole volume of Baltic waters (Snoeijs-Leijonmalm et al. 2017; Uścinowicz 2011). As a result, all compounds which entered the Baltic Sea are circulating for a very long time. This is especially important in case of hazardous substances and nutrients responsible for eutrophication of the Baltic Sea. Furthermore, the oxygen depletion increasing yearly on the Baltic Sea bottom depends on the oceanic waters inflows, affecting biota and chemical processes. Exceptions are visible after Major Baltic Inflows (MBI) which occur intermittently (ca. once every 10 years) when fresh oceanic water enters the Baltic Sea. Due to their high salinity and therefore density, oxygenated oceanic waters cover the bottom of central Baltic Sea, improving the environmental conditions in this area. On the other hand, the conditions in Gulf of Finland are getting worse, as the anoxic water masses are pushed into this region. Even though the MBI locally increases oxygen levels, the changes of environments oxy-redox potential determine the dynamics and directions of chemical processes

Baltic ecosystem degrades because of human activity since the 20th century—the major threats are eutrophication, pollution and overfishing. The inflow of saline water is very limited in comparison with the impact of the drainage area—nearly 200 rivers enter the Baltic Sea. 28 major rivers cover up to 80% of the drainage area, and 12 of those 28 are classified as eutrophic, responsible for the transport of nutrients estimated on total annual input of more than 410,000 tonnes of nitrogen and almost 21,000 tonnes of phosphorus (both natural origin and resulting from agricultural use on land) (Snoeijs-Leijonmalm et al. 2017).

It is estimated that nearly 85 millions of people live in the drainage area and almost 18% of them lives within 10 km from the shore (Snoeijs-Leijonmalm et al. 2017). While nearly 80% of the Scandinavian peninsula is covered by forests,

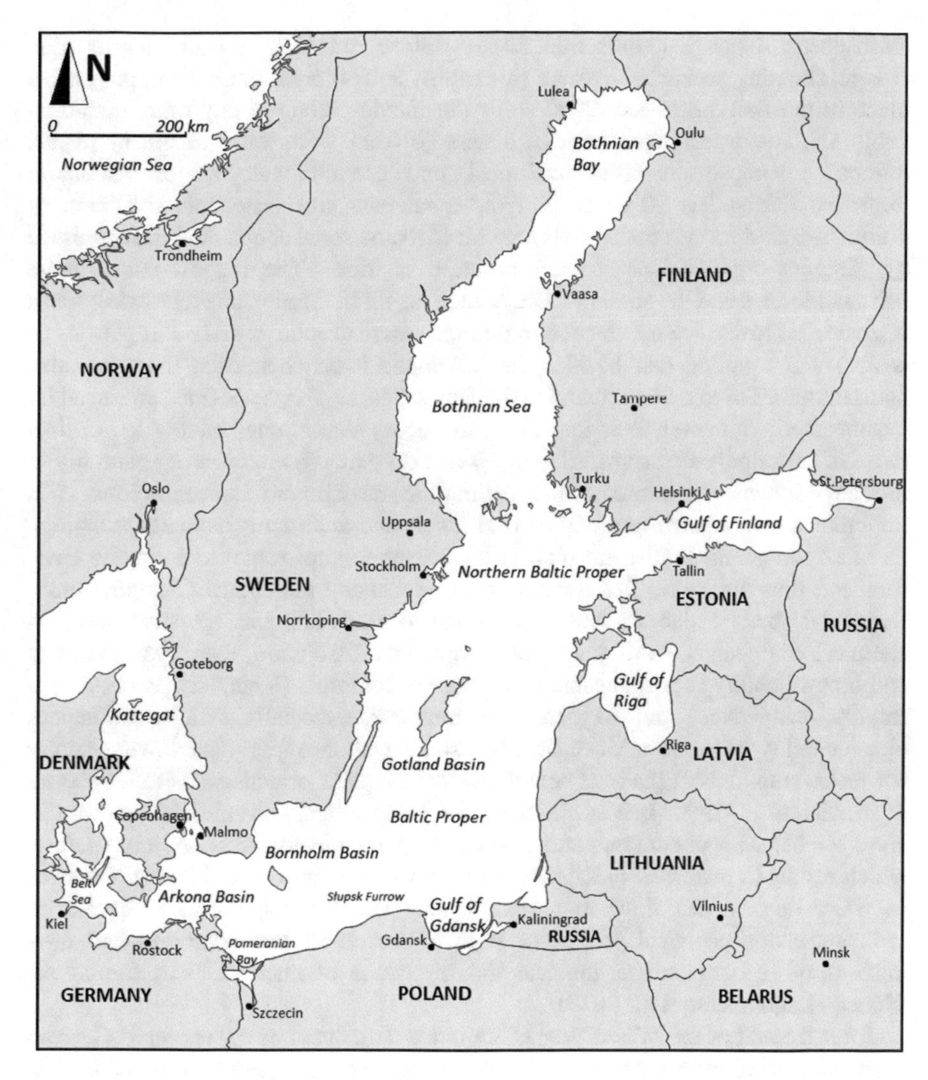

Fig. 1 Baltic Sea location and subdivision

eastern and southern Baltic countries can be characterized by extensive agriculture on ca. 65% of their area. Excessive use of fertilizers, large farms with extensive livestock production, fossil fuel burning, discharge from wastewaters, industrial point sources, fish farms and deforestation (forests act as natural nutrient traps) are the main anthropogenic sources of nutrients and contaminants to the Baltic Sea (Snoeijs-Leijonmalm et al. 2017).

Along with eutrophication, chemical pollution is one of the major threats to the Baltic Sea ecosystem. Contaminants enter the Baltic Sea from various sources:

atmospheric deposition, industrial waste, diffuse runoff from land, riverine discharge, shipping accidents, leakage from ships hulls (very intensive shipping takes place in the Baltic Sea—ca. 2000 ships are moving along at any time, including cargo ships and passenger ferries), discharge from wastewater treatment plants, dumped chemicals including chemical and conventional weapon. Situation improved during last 20 years with the awareness and systematic reduction of pollutant emissions, yet their levels are still high and some above the limits assessed by different organizations for non-polluted matrices. The biggest concern are organochlorines and the so-called heavy metals, yet in recent years a growing focus is given also to surface-treatment compounds, pharmaceuticals and microplastics. It is also worth noting that besides the manmade toxic chemicals, there are also natural sources of toxic compound in the Baltic Sea, e.g. hepatotoxines produced by cyanobacteria, however their impact on the ecosystem is uncountably lower than those of anthropogenic origin. Also, most of heavy metals are occuring naturally in the environment; therefore, while estimating background concentrations it is important to distinguish between natural levels and anthropogenic contamination.

Most contaminants that are bind to sediments are not removed from the environment, thus the Baltic Sea became a sink for many pollutants. Changing environmental factors, due to MBI, drive the degradation and transformation of chemical compounds, as they control compounds dissolution, their cycle, toxicity and bioavailability for organisms. Main factors controlling chemical processes are salinity, temperature and oxygen concentrations, especially salinity influences bioavailability of some toxic elements—some authors report that lower salinity correlates with higher uptake of heavy metals by aquatic organisms (McLusky et al. 1986; Hall et al. 1995; Lee et al. 1998). Ongoing climate change also may influences the behavior of contaminants as they cause notably higher water temperature, which results in increased precipitation and lower salinity (BACC 2015). Changes in water density and thus stratification may affect resuspension processes of sediment-bound chemical pollutants, temperature determines solubility and fluctuations of redox potential indicate the directions of chemical transformations (Snoeijs-Leijonmalm et al. 2017).

After the end of the World War II, significant amounts of the captured German chemical arsenal were dumped to the Baltic Sea. The dumping took place in designated deep areas. Bornholm Deep and Gotland Deep were chosen for this purpose due to their depth and bottom currents which will prevent the munitions to allocate. According to HELCOM MUNI report (Knobloch et al. 2013) 10,000–12,000 tonnes of chemical munitions could have been dumped in Polish Exclusive Economic Zone (EEZ). The total quantity of Chemical Warfare Agents (CWA) contained in munitions dumped in the Baltic Sea region is estimated to reach over 30,000 tonnes, and about 31% of that is considered to be arsenic-containing chemicals (arsine oil, Clark I, adamsite) (MERCW 2006; Paka and Spiridonov 2001). But as established later, munitions were scattered in and outside of those areas (Bełdowski et al. 2016; Kasperek 1997; HELCOM 1995).

1.2 Arsenic

Arsenic is a chemical element with an atomic number 33 and atomic mass of 75. It occurs naturally in the environment; therefore it is present in all of its compartments. However, nowadays it is estimated that nearly 70% of arsenic in the environment origins from anthropogenic activities like the use of pesticides, coal burning, mining and smelting processes. US EPA reports that one of the most important sources of arsenic in marine environment is vessel bilgewater discharges (along with other pollutants including pathogenic bacteria) (US EPA 2010). Also offshore and gas industry can be a source of arsenic from drilling muds (Tornero and Hanke 2016).

Arsenic is widely distributed in the environment and, depending on the oxy-redox conditions, it occurs in the environment in four oxidation states (As3−, As0, As3+, As5+). All arsenic minerals and compounds are soluble but the ability of migrating into the hydrosphere is restricted with binding by argillaceous minerals, iron and manganese hydroxides or organic matter (Cullen and Reimer 1989). Specific bacteria, some algae species and fungi occurring in marine environment have the ability of binding and methylation of inorganic arsenic. Methylarsenic acid (MMA) and dimethylarsinic acid (DMA) are significantly less toxic than inorganic arsenic compounds. Marine samples of different origin are thought to comprise the greatest number of arsenic compounds occurring naturally and are reported to contain concentrations of arsenic even one order magnitude higher than comparable samples from terrestrial environments (Flora 2015; Reimann et al. 2009). Concentrations of arsenic in natural waters depend mainly on the bedrock composition (as chemical and mechanical weathering is considered to be the main natural source of this element) and occurrence of external contamination sources.

In marine environment, arsenic occurs in different oxidation states, both in organic and inorganic compounds. Organic arsenic species are assumed to be less toxic than inorganic, and arsenic (III) compounds are the most toxic of all arsenic species present in the environment, recursively more toxic than As(V) (Cullen and Reimer 1989). Under reducing conditions, arsenic (III) is the dominant species while arsenic (V) is generally stable in oxygenated environments. Arsenate [As(III)] occur in anaerobic conditions probably as a result of the competition between Fe (III) and As(V) as an electron acceptor for microbial respiration or the As(V) oxidation by manganese oxides. For this reason, changes in climate conditions can influence biogeochemical cycles, which could result in reemission of heavy metals deposited in sediments, including arsenic. Eutrophication of the Baltic Sea may increase due to elongated growth season, which causes oxygen depletion in bottom waters. Anaerobic conditions foster the dissolution of iron and manganese oxides and oxyhydroxides, and release of adsorbed arsenic in its most toxic form into the water column, where it can be assimilated by organisms (Yu et al. 2016; Fauser et al. 2013; Bissen and Frimmel 2003). Cartensen et al. (2014) report a constant extension of areas affected with hypoxic conditions on the bottom of the Baltic Sea. Eh values determine oxidation or reduction of arsenic species, while type and chemical composition of sediments is related to arsenic bounding with sulfur, which causes its immobilization. Absorption on iron and manganese oxyhydroxides influences the cycle of arsenic between sediments and water column (Bissen and Frimmel 2003; Cullen and Reimer 1989).

Arsenic is assumed to be toxic to plants, animals and human and is considered a potential carcinogen. It disrupts enzymatic processes in cells, causes cell walls breakdown, inhibits mitochondria functions, affects proteins formation by its high affinity to sulfhydryl groups, inhibits phosphate insertion to DNA, affecting transmission of genetic information (Harkabusova et al. 2009; Bissen and Frimmel 2003; Nicholas et al. 2003; Niedzielski et al. 2000). Chronic exposure to elevated arsenic concentrations can cause disturbance in nervous system and heart diseases. It is also assumed that bladder and lung cancer may be caused by chronic arsenic poisoning (Flora 2015). Seafood is considered one of the main sources of arsenic in human diet (Flora 2015); therefore, investigating the marine environment might be crucial from the perspective of fish consumers. Regular research performed by National Sea Fisheries Institute in Gdynia (Polak-Juszczak 2009; Usydus et al. 2009) shows that arsenic levels in muscle tissue of fish available on polish market are not elevated and do not pose a risk for human health. Despite that there is no direct threat for the fish consumers, the environment should be monitored due to changes undergoing in the Baltic Sea which could potentially affect bioavailability of arsenic species for marine organisms.

In the Baltic Sea a potential threat of arsenic contamination in time is related to chemical munition dumping, as some of the CWA consist of toxic organoarsenic compounds. The arsenic-containing CWA are able to contaminate bottom sediments due to their low solubility. According to the MERCW project investigations (Modeling of Ecological Risks Related to Sea-Dumped Chemical Weapons: www.mercw.org), elevated arsenic levels were found in bottom sediments out of the primary dumping site zone, while studies performed within the CHEMSEA project (Chemical Munition Search and Assesment: www.chemsea.eu) indicated the presence of degradation products of arsenic containing chemical warfare agents as far as 40 m away from the dumped munitions. Concentrations in dumping sites are significantly higher than concentrations measured in areas where no chemical munitions were dumped (Bełdowski 2016a).

2 Arsenic in the Baltic Sea

In the area of most countries surrounding the Baltic Sea, levels of arsenic in top soil and stream sediment are low and do not exceed 7 $\mu g\ g^{-1}$. This part of northern Europe can be characterized by the lowest concentrations of arsenic in soils on the whole continent. Only the area of Northern Sweden stands out of this trend, exceeding 9 $\mu g\ g^{-1}$ in subsoil, topsoil and reaching up to 36 $\mu g\ g^{-1}$ in stream sediments, as it stays under the strong influence of mining (Uścinowicz 2011; Salminen 2005). Catchment properties are directly reflected in marine samples.

2.1 Water

There is not much research on arsenic concentrations in the Baltic Sea water; however, the first reports came in the early 1980s when the analytical technique of Hydride Generation was combined with Atomic Absorption Spectrometry and was successfully implemented to environmental sciences.

In 1982 and 1983, Stoeppler et al. (1986) conducted a comparative study on water and algae samples from the Nort Sea and Baltic Sea. They reported dissolved arsenic in marine water with a mean value of 0.76 $\mu g\ l^{-1}$ (range from 0.45 to 1.11 $\mu g\ l^{-1}$) for 17 stations from the Baltic Sea, which were low comparing to the values obtained for the samples from the North Sea (2–3.1 $\mu g\ l^{-1}$). The difference in arsenic values between filtered and non filtered samples was practically negligible in the case of Baltic waters. Authors underline that no clear seasonal variations were observed, yet their conclusion was driven after sampling only on two different days in 1982. Within the MERCW project, Khalikov and Savin (2011) measured arsenic concentrations in bottom waters collected in the area of chemical munition dumpsite in the Bornholm Deep. They report mean concentrations of 0.63 $\mu g\ l^{-1}$ in water from the dumpsite and 0.55 $\mu g\ l^{-1}$ beyond the dumpsite. Also the water collected from above of the sediment was one order of magnitude lower than pore waters extracted from the same sediment. Andreae and Froelish (1984) report the values of total As in Baltic Sea water on the level of 0.6 $\mu g\ l^{-1}$. The highest arsenic concentrations, reaching up to even 510 $\mu g\ l^{-1}$, were measured in water entering the Bothnian Bay close to the smelter on the shore of Sweden (Blanck et al. 1989).

There is hardly more reports on arsenic in water of the Baltic Sea; however, it may be an important issue, as in anoxic conditions arsenic usually occurs in the most toxic species As(III) (Maher and Butler 1988). Due to the increase of anoxic and hypoxic areas in the Baltic Sea, arsenic might be released from the sediments and enter the biogeochemical cycle in its most harmful form.

2.2 Sediments

Total arsenic concentration levels in Baltic Sea bottom sediments are highly variable and depend on the distance from arsenic source and its type, as well as on the type of sediments covering the bottom and environmental conditions (Emelyanov 2007). The fraction of sediment grains <2 μm is one of the major sinks for pollutants introduced into natural waters, mostly because of the abundance of clay minerals, together with associated coatings of organic material and the presence of iron and manganese (oxy-)hydroxides (Emelyanov 2007; Garnaga et al. 2006). Therefore, sediment characteristics have an important influence on arsenic enrichment and the spatial variability of arsenic concentrations depends largely on the sedimentary environment, sediment mineralogy and redox conditions, as well as the amount of organic matter and the argillaceous fraction contents in the sediments (Bełdowski et al. 2016a). Arsenic concentrations in Baltic Sea sediments vary spatially over a very wide range, from 0.3 to 277 $\mu g\ g^{-1}$ (Uścinowicz 2011).

2.2.1 Bornholm Deep

Maximum concentrations of arsenic measured in the Bornholm Basin in an unpolluted area reach 29 $\mu g\ g^{-1}$. Part of the Bornholm Basin close to northern border of Polish EEZ is one of the most arsenic contaminated areas of the Baltic Sea (Uścinowicz 2011; MERCW 2006). One of the hot spots characterized by elevated concentrations of arsenic is situated in close vicinity to the chemical weapon dumpsite in Bornholm Deep (Missiaen et al. 2010). Based on normalized As/Fe concentrations there also appears to be a strong influence of fishing harbors (Uścinowicz 2011).

The Bornholm Deep is assumed as a sink for many pollutants. Khalikov and Savin (2011) report arsenic concentrations in sediments from Bornholm Deep munition dumpsite in a wide range between 8 and 210 $\mu g\ g^{-1}$, with the highest values in the layer of the top 5 cm of sediment cores. Samples with the highest arsenic values were collected very close to the objects lying on the sea floor and close to a sunken wreck. Mean concentration of arsenic in pore waters from the collected sediments was 4 $\mu g\ l^{-1}$ and when distinguishing between the center of dumping area and area beyond—the mean values were 27.4 and 2.8 $\mu g\ l^{-1}$, respectively. Arsenic concentrations measured in bottom water, above the sediment was one order of magnitude lower than in pore waters. In this area, Sanderson et al. (2008) measured the highest arsenic concentration (210 $\mu g\ g^{-1}$) in fine grained pelitic mud sample. Also Emelyanov et al. (2010) found up to 210 and 277 $\mu g\ g^{-1}$ of arsenic in fine-grained sands, pelitic clayey mud and in aleuro-pelitic mud sediments. In the most recent study performed in this region a mean concentration of only 17 $\mu g\ g^{-1}$ was reported for sediments from the area of Bornholm Deep (Bełdowski et al. 2016a).

2.2.2 Southern Baltic Proper

In the sediments of the southern Baltic Sea, arsenic occurs in concentrations ranging from less than 5 $\mu g\ g^{-1}$ in sandy sediments, in muddy-sands they reach the values between 5 and 15 $\mu g\ g^{-1}$ and reach up to 29 $\mu g\ g^{-1}$ in muds (Uścinowicz 2011). The lowest concentrations were measured in the area of Słupsk Furrow and in Puck Bay, the inner part of the Gulf of Gdańsk. In the sediments of the western Gulf of Gdańsk the concentrations of arsenic range from 5 to 12 $\mu g\ g^{-1}$, depending on the type of bottom sediments. Szczepańska and Uścinowicz (1994) also reported arsenic concentrations ranging from 2 to 9 $\mu g\ g^{-1}$ in the area of central Polish coast and between 10 and 15 $\mu g\ g^{-1}$ in the Słupsk Furrow. In the surface layer of aleurite-clays concentrations are higher—up to 15–29 $\mu g\ g^{-1}$.

Several studies were conducted during last few years in the area of Polish Exclusive Economic Zone, concerning the suitability of Słupsk Furrow for creating a wind farm and due to verify the possibility of dumping chemical weapon in the Gdańsk Deep after the II World War.

The research in Słupsk Furrow was performed in the central part of the Baltic Sea in Polish EEZ (from the longitude of the city Kołobrzeg to the town of Łeba). Measured arsenic concentrations varied from 0.3 to 22.7 μg g^{-1} and relatively low correlation factor values between arsenic and iron ($r = 0.24$) and between arsenic and organic matter contents ($r = 0.49$) were noted. In cores of sediments arsenic concentrations were rather stable up to the sediment layers dated for the year 1815. No enrichment after the industrial revolution (ca. 1900) was observed, what may indicate that values exceeding the geochemical background can be still assumed as natural. The highest arsenic values were observed close to the Bornholm Deep which is a sink of pollutants binding to the fine sediments characteristic for this deep region (Zaborska et al. 2016; Szczepańska and Uścinowicz 1994). Zaborska et al. (2016) report only slight exceed of the value of 13 μg g^{-1}—the geochemical background concentration given by Uścinowicz (2011). Bełdowski et al. (2016a) also report low values of arsenic concentrations in sandy sediments of Słupsk Furrow (mean value 5.7 μg g^{-1}) and in the region located north from the Rozewie Cape—5.1 μg g^{-1}.

Szczepańska and Uścinowicz (1994) report arsenic concentrations in Gdańsk Basin slightly exceeding 20 μg g^{-1}. Because of chemical munition dumping, Gdańsk Deep was studied in details during several research projects (MERCW, CHEMSEA, MODUM). Emelyanov (2007) observed concentrations of arsenic in Gdańsk Basin and Gotland Basin reaching 70 μg g^{-1}; however, in recent studies such high values were not reported. Bełdowski et al. (2016a) report mean As concentrations of 15.5 μg g^{-1} in muddy sands collected in the vicinity of dumping areas (95 m depth, higher contents of sediments fine fraction) and a mean value of 9.8 μg g^{-1} in the area of the Gulf of Gdańsk (within 3–10 km from the Polish coast, 52 m depth, higher contents of sandy sediment fraction).

2.2.3 Gotland Deep and Lithuanian EEZ

Several studies concerning the influence of chemical munition on the Baltic ecosystem were performed also in the Gotland Deep and in the nearby area, spreading to the Lithuanian coast.

Garnaga and Stankevičius (2005) and Garnaga et al. (2006) performed sediments sampling along the Lithuanian shore and along a transect determined to the Gotland Deep munition dumpsite. Measured arsenic concentrations ranged from 1.1 to 19 μg g^{-1}. In the samples collected in the shallow areas (depth between 11 and 46 m) in the vicinity of 10 NM from the shore the concentrations did not exceed 4 μg g^{-1} of arsenic, with average of 2.2 μg g^{-1}. Slightly further from the shore (11–57 NM) and in deeper areas (46–75 m) arsenic concentrations were still around 2 μg g^{-1} despite the different sediment type. In the deep area of the dumpsite (depth between 85–125 m) the average As concentrations reached 10.9 μg g^{-1}. Authors point out two possible reasons for those higher concentrations —it may be the impact of chemical munitions and higher contents of small fraction of sediments, which is characteristic for this deep area of the Gotland Basin. According to the study, the directions and velocities of bottom currents could inhibit the spreading of contaminants from the dumpsite. Also an interesting pattern of arsenic to iron ratio was observed in the samples from chemical munition dumping site—the correlation factor between arsenic and iron are significantly lower than in the samples collected outside the dumping area (Garnaga et al. 2006). Similar conclusions were drawn by Bełdowski et al. (2016a), who point out a different behavior of arsenic in dumpsite areas in comparison to reference areas, basing on the correlation between concentrations of arsenic and such parameters as organic matter contents, sediment grain size, iron and manganese concentrations in samples with and without detected arsenic based CWA. In their study Bełdowski et al. (2016a) reported mean total arsenic concentration on the level of 13 μg g^{-1} in the sediments form the dumpsite and mean arsenic concentration 6.2 μg g^{-1}outside the primary dumpsite ca. 67 NM from the Lithuanian coast. However these concentrations are relatively low compared to those reported in other Baltic Sea regions and North Sea investigations (Bełdowski et al. 2016a; Knobloch et al. 2013; Garnaga et al. 2006).

2.2.4 Bothnian Sea

In the Bothnian Sea—the highest concentrations in aleurite-clay were observed in the range from 167 to 216 μg g^{-1} (Uścinowicz 2011).

Bothnian Bay is one of the most polluted areas of the Baltic Sea and arsenic is assumed to be the biggest threat of all pollutants in this area. Arsenic loads to the Bothnian Bay sediments origin from the Rönnskärsverket smelter in Sweden (located on the western shore of the Bothnian Bay), and may be harmful to marine organisms (Borg and Johnson 1996; Vallius 2014). Arsenic concentrations were increasing at this location since the middle 1970s reaching levels of 50 μg g^{-1} at that time. Arsenic concentrations were still increasing during the 1980s and in the 1990s reached up to 100 μg g^{-1}. Even though the cessation of smelter activity arsenic levels declined only slightly after reaching this maximum and still remain elevated comparing to other areas.

Widerlund and Ingri (1995) conducted a complex study in the region of Kalix river estuary in northern Sweden, which stays under the strong influence of a Skellefteå sulphide ore smelter. Concentrations of arsenic in top 1 cm of the sediment cores ranged from 38 to 44 μg g^{-1} while the maximum concentrations were observed at the depths of 4 to 12 cm and ranged from 160 to 170 μg g^{-1}. Those values are much higher than observed in other areas of the Baltic Sea. Also in pore water samples the highest concentrations were observed downward the core—even 70–90 times higher at the depths of 10–15 cm than in the surface 1 cm. Additionally a measurement was made in sediments collected in the Kalix river basin ca. 40 km up from the estuary—arsenic concentrations in river sediments valued 5 μg g^{-1} and in riverine water 0.16 μg l^{-1}.

Patterns of arsenic distribution in the cores of sediments reflect the processes occurring in the sampling region. Redox conditions at the boundary layer between sediments and bottom water may induce if the particular area will behave as a sink or a source of arsenic in the environment. Different pattern for arsenic in sediments and in pore waters may indicate that different processes were occurring in time—remobilization of As due to reduction and dissolution of iron (III) oxides and manganese-oxides, removal of As from sediment to pore waters or opposite, probably due to scavenging of arsenic by sulphide phase. Basing on the estimations of arsenic fluxes, the authors suggest that such high levels of arsenic in pore waters may indicate that this element in majority comes from an inner source and not from terrestrial sources, and could result from former deposits released from the sediment and circulating through the sediment (Widerlund and Ingri 1995).

2.2.5 Gulf of Finland

Several studies on arsenic concentrations in sediments were conducted in the Gulf of Finland. Since the mid-1990s, Vallius with coworkers, have been investigating the state of heavy metals pollution in this area and evaluating the ongoing changes in arsenic concentrations (Vallius 1997).

In 1998 and 1999 Vallius and his coworkers published the results obtained for the region of Gogland Island and eastward open sea. Salinity of this area varies from 0–5 on the surface to 5–8 in bottom waters, due to relatively high input of water via the Neva river. Neva stays under a high impact from industrial areas and commercial harbours, and therefore it is a principal carrier of pollutants to the Gulf of Finland, as riverine input is one of the most important sources of contaminants to the sea. The Coriolis force establishes the water current in this area in such way that the transport of the pollutants is driven towards the north-east.

The authors observed minimum and maximum concentrations of arsenic on the level of 4 and 27 μg g^{-1}, respectively. The highest concentrations, close to or exceeding 20 μg g^{-1}, were found in the uppermost sediments layers (0–20 cm), close to the coasts of Estonia. Authors underline that those results correlate with information from geochemical maps showing anomaly arsenic concentrations in top soil in Dictyonema shale—in this particular area there is a probable impact form Sillamäe oil shale electrical power plant emissions. Westward the concentrations in surface sediments are decreasing, however are elevated and reach 25 μg g^{-1} in the sediment layers below 60 cm (Vallius and Leivuori 1999; Vallius and Lehto 1998).

Later studies bring comparable arsenic concentrations results for sediments in this area (Yli-Hemminki et al. 2016; Vallius 2012, 2015; HELCOM 2010; Vallius et al. 2007). Minimum arsenic concentration of 7.72 μg g^{-1} and maximum of 19.1 μg g^{-1} (mean value: 12.49 μg g^{-1}) in the surface sediments correspond to previous studies. In deeper layers (8–15 cm), slightly higher arsenic concentrations were observed (minimum 9.24 μg g^{-1}, maximum 25.4 μg g^{-1}, mean value

13.3 μg g^{-1}), which may indicate that the emission of arsenic to the marine environment of Gulf of Finland decreased, as the deeper layers reflect elder sedimentary material (Vallius et al. 2007). The highest arsenic concentrations within the Gulf of Finland were measured in samples collected close to the Russian-Finland border and ranged from 22.1 to 27.8 μg g^{-1} (Vallius 2012).

2.2.6 Skagerrak

Although Skagerrak is not a part of the Baltic Sea it is an important strait, connecting the Baltic Sea with the North Sea., and requires a comment. Tørnes et al. (2002) conducted a study on arsenic and arsenic containing CWA in the area of Skagerrak, where shipwrecks with chemical munitions were sunk. Authors reported a background concentrations of arsenic, ranging from 42.8 to 49.3 μg g^{-1}, however in four samples an elevated arsenic concentration was noted, reaching from 75 μg g^{-1} up to even 480 μg g^{-1} in the samples where arsenic-containing CWA (Clark I, triphenylarsine and bis(diphenylarsine)oxide) were also detected.

Summarizing, arsenic concentrations in the Gulf of Finland are considered to be rather low in comparison to other Baltic Sea areas. Elevated values are observed in deeper layers of the sediments, thus only a few species of benthic biota dwell deeper than 20 cm; therefore, arsenic should not be a threat to marine biota inhabiting this area (Vallius 2014, 2015). However, Yli-Hemminki et al. (2016) suggest that in the oscillating oxygen conditions in eutrophic Gulf of Finland, deposits of iron-manganese concretions forming a large storage of arsenic and phosphorus may be an inner source of arsenic returning to the biogeochemical cycle as the concretions may dissolve through microbial reactions in anoxic conditions. Results obtained in research on arsenic in Baltic Sea sediments and other matrices are given in Table 1.

2.3 Organisms

As seafood is considered to be one of the major arsenic sources in human diet, investigating marine organisms gains researchers interest, even though it is believed that non-toxic arseobetaine dominates in marine food prodcts (Flora 2015). The number of arsenic species identified in marine organisms is estimated on over fifty (Francesconi 2010) and probably could be even higher as a result of the large number of lipophilic arsenic compounds reported in the last few years (Francesconi and Schwerdtle 2016). Rahman et al. (2012), in their publication about arsenic in aquatic food chains indicated that significant differences occur between predominating arsenic species in organisms on the same trophic level from freshwater environments and from marine environments. It is also reported that increased arsenic concentrations occur together with increased salinity (Flora 2015; Kuenstl et al. 2009). Moreira et al. (2018) conducted a study on arsenic toxicity for Pacific

Table 1 Arsenic concentrations in marine samples of different type and origin and analytical methods used in reference literature (Authors name in alphabetical order)

Matrix, location	Analytical method	Reference	As concentrations
Water	HG-AAS with quartz tube	Andreae and Froelich (1984)	0.6 μg l^{-1}
Cod and herring	HG-AAS	Ballin et al. (1994)	0.56–0.87 μg g^{-1}
Sediments from chemical munition dumpsites	HG-AAS	Bełdowski et al. (2016a)	Sands: 0.3 μg g^{-1} Muds: 23 μg g^{-1}
Water	HG-AAS	Blanck et al. (1989)	510 μg l^{-1}
Sediments	AAS with electrothermal atomizer	Emelyanov (2007), Emelyanov et al. (2010)	210–277 μg g^{-1}
Muddy sediments, Bornholm Deep	ICP-MS	Fausner et al. (2013), Sanderson et al. (2008)	Up to 210 μg g^{-1}
Sediments, Gotland Deep, Lithuanian EEZ	Electrothermal AAS	Garnaga et al. (2006), Garnaga and Stankevičius (2005)	1.1–19 μg g^{-1}
Blue mussels, Baltic Sea and Baltic Sea	INAA	Karbe et al. (1977)	3.6–5.0 μg g^{-1}
Bottom waters and sediments from chemical munition dumpsite (Bornholm Deep)	XRF	Khalikov and Savin (2011)	Water: 0.55–0.63 μg l^{-1} Sediments: 8–210 μg g^{-1}
Fish, Scandinavian part of Baltic, Belt Sea	*x*	Larsen and Francesconi (2003)	0.66–1.0 μg g^{-1}
Fish and water, Bothnian Bay	INAA	Ljunggren et al. (1971)	Herring: 0.14–1.5 μg g^{-1}
Sediments from chemical munition dumpsites	XRF	MERCW (2006)	
Organisms, Bothnian Bay, Northern Baltic Proper	ICP-MS	Nfon et al. (2009)	Plankton: 0.36 μg g^{-1} Zooplankton: 0.62 μg g^{-1} *Mysidae*: 0.79 μg g^{-1} Herring: 0.64 μg g^{-1}
Fish, polish coastal zone	HG-AAS	Polak-Juszczak (2013)	~1.2 μg g^{-1}

(continued)

Table 1 (continued)

Matrix, location	Analytical method	Reference	As concentrations
Marine sediments, subsoil, topsoil, stream sediments, stream waters, whole Europe	XRF ICP-MS	Salminen (2005)	Foodlplain sediments: 1–390 μg g^{-1} Subsoil: 0.2–594 μg g^{-1} Topsoil: 0.3–280 μg g^{-1} Stream sediments: 1–241 μg g^{-1} Stream water: 0.01–27.3 μg l^{-1}
Water and brown algae, Baltic Sea and North Sea	*x*	Stoeppler et al. (1986)	Water: 0.76 μg l^{-1} *F. vesiculosus*: 40 μg g^{-1}
Sediments, Skagerrak	ICP-MS	Tørnes et al. (2002)	42.8–480 μg g^{-1}
Blue mussels	ICP-MS	Turja et al. (2014, 2015)	5.6–13 μg g^{-1}
Sedimets, whole Baltic Sea	ICP-AES	Uścinowicz (2011)	0.3–277 μg g^{-1}
Sediments, Gulf of Finland	ICP-AES	Vallius et al. (2007); Vallius (2012, 2014, 2015)	7.72–27.8 μg g^{-1}
Sediments, Kalix River estuary (northern Sweden)	ICP-MS	Widerlund and Ingri (1995)	38–170 μg g^{-1}
Sediments, Słupsk Furrow	HG-AAS	Zaborska et al. (2016)	0.22–22.66 μg g^{-1}

Abbreviations used in the table: **HG-AAS**—Hydride Generation Atomic Absorption Spectrometry, **AAS**—Atomic Absorption Spectrometry, **ICP-MS**—Inductively Coupled Plasma Mass Spectrometry, **XRF**—X-Ray Fluorescence spectrometry, **ICP-AES**—Inductively Coupled Plasma Atomic Emission Spectroscopy, **INAA**—Instrumental Neutron Activation Analysis

oyster (*Crassostrea gigas*) and report arsenic EC50 (Effect Concentration visible in 50% of studied population) values ranging from 0.002 to 219 μg l^{-1}. Authors underline that As toxicity depends on environmental conditions and matrix properties.

In 2009 Nfon et al. published the results of a complex study on total arsenic and other trace elements concentrations in Baltic organisms from different trophic levels—plankton, zooplankton, crustaceans from *Mysis* sp. and herrings from Northern Baltic Proper and Bothnian Bay. The average concentrations of arsenic were 0.36 μg g^{-1} for plankton, 0.62 μg g^{-1} for zooplankton, 0.79 μg g^{-1} for *mysidae* and 0.64 μg g^{-1} for herring (dry weight). The obtained results were at very low levels and did not differ significantly between the species. Basing on this study it is impossible to conclude that biomagnification of arsenic takes place in Baltic Sea environment. Aquatic species do bioconcentrate and bioaccumulate arsenic; however, it is assumed to be relatively quickly excreted from the organism so the possibility of accumultion via trophic transfer is quite likely (Nfon et al. 2009).

2.3.1 Algae

One of the first reports on arsenic concentrations in Baltic species refers to brown algae. *Fucus vesiculosus* was chosen as a dominant species in the Baltic Sea, and easily available for sampling (Stoeppler et al. 1986). Authors investigated different sites on the German coast of Baltic Sea and North Sea for comparison. Total arsenic levels reached up to 40 μg g^{-1} dry weight of algae, and 95% of that was assumed as chemically stable organoarsenic compounds. Authors report a clear seasonal variation in arsenic levels with the highest values in winter and lowest in summer, when the vegetation period begins and the plant material is formed. Additionally arsenic concentrations were investigated in other algae species from the Baltic Sea and the Mediterranean Sea. These indicate that there are significant species and habitat-specific variations in arsenic accumulation and that the brown seaweed *Laminaria saccharina* from the Baltic Sea is probably a more effective arsenic accumulator (Stoeppler et al. 1986).

Blanck et al. (1989) underline the significance of algae investigations in the context of arsenic measurements. Due to some chemical similarities of arsenic to phosphorus, those two elements may compete to be bound in organisms; therefore, the uptake of arsenic by algae might be the most significant route of transferring this element into aquatic biota. Algae have a high capacity to transform arsenate for example to arsenite, methylated arsenic acids, arsenosugars and arsenolipids, which makes algal metabolism crucial for the distribution of arsenic in the ecosystem. No biomagnification of arsenic occurs, but several organic arsenic compounds formed by algae, and their degradation products, have been identified at higher trophic levels of the ecosystem. Algae are also the main target of arsenate toxicity (Blanck et al. 1989; Nfon et al. 2009; Kuenstl et al. 2009; Rahman et al. 2012).

2.3.2 Mussels

In 1977 a first study on arsenic concentrations on mussels from the Baltic Sea was performed. Along the German coasts of the North Sea and Baltic Sea blue mussels (*Mytilus edulis*) were sampled and a comparative study on trace elements concentrations was conducted. Mean concentrations of arsenic from mussels from the Baltic Sea was 4.1 μg g^{-1} (range 3.6–5.0 μg g^{-1}) while from different sites in the North Sea coast it differed from 7.4 to 9.9 and 12.1 μg g^{-1} in vicinity to Helgoland (Karbe et al. 1977).

More recently, mussels were used for estimation of pollution in different sites impacted by anthropogenic activity. Turja et al. (2014) conducted a caging experiment using *Mytilus trossulus*. Organisms collected in a reference area were placed for three months in cages situated in three locations, including a contaminated zone of Swedish coast, within the Bothnian Bay. Despite different conditions of the environments and different level of contamination at all three sites, all organisms reflected similar arsenic concentration values ranging from 8.6 to 13 $\mu g\ g^{-1}$ (Turja et al. 2014). Similar caging experiment was performed for the waters of Gulf of Finland, south of the city Helsinki. No significant differences were found in the used mussels, and the arsenic concentrations in their tissue ranged from 5.6 $\mu g\ g^{-1}$ on the first day to 6.4 $\mu g\ g^{-1}$ by the end of the experiment (Turja et al. 2015).

2.3.3 Fish

Parts of Scandinavia and the North Baltics are considered as consumers of the highest amounts of pelagic and demersal fish (Sioen et al. 2009; Taylor et al. 2017). Several studies and monitoring are performed in the area of Baltic Sea to investigate the quality of fish and potential threat for consumers, connected with different contaminants in fish meat.

One of the earliest research on arsenic concentrations in fish [perch (*Perca fluviatilis*) and pike (*Esox lucius*)] was conducted in an uncontaminated area of the brackish-waters of Bothnian Bay. Total arsenic concentrations in those fish ranged between 0.14 and 0.28 $\mu g\ g^{-1}$ dry weight in muscle tissue and concentrations in liver ranged from 0.24 to 1.5 $\mu g\ g^{-1}$. Higher concentrations of contaminants in liver are observed as a result of the purifying function of this organ, which binds and cumulates toxic compounds. Baltic herring from the northernmost part of the Bothnian Bay was reported to contain 1.21 $\mu g\ g^{-1}$ of arsenic in a muscle tissue. Fish from the Swedish east coast generally contained five to ten times higher total arsenic concentrations, compared to the same species caught in other regions of the Baltic Sea, probably due to the heavy contamination of the Bothnian Bay (Ljunggren et al. 1971).

In 1994 Ballin et al. (1994) reported mean concentrations of arsenic in Baltic herring on the level of 0.87 and 0.56 $\mu g\ g^{-1}$ wet weight in cod muscle, which were lower than the concentrations in fish from the North Sea, Shetlands and Greenland.

Larsen and Francesconi (2003) measured arsenic concentrations in muscle of three species of fish (cod—*Gadus morrhua*, herring—*Clupea harengus* and flounder—*Platychthis flesus*) from Scandinavian parts of Baltic—Danish waters, Belt Sea, Kattegat and from North Sea for comparison. In fish from Baltic and Belt Sea (with salinity 8 and 12 respectively) the concentrations of arsenic in all three fish species did not exceed the value of 1 $\mu g\ g^{-1}$ g wet weight (average values 0.77–0.98 $\mu g\ g^{-1}$ in herring, 0.66–1.0 $\mu g\ g^{-1}$ in cod and 0.89 $\mu g\ g^{-1}$ in flounder), while the concentrations in fish from more saline Kattegat (salinity 28) were significantly higher (1.71 $\mu g\ g^{-1}$ for herring, 4.77 $\mu g\ g^{-1}$ for cod, and 2.72 $\mu g\ g^{-1}$ for flounder). In comparison with the fish from the North Sea (salinity 32), the levels from Kattegat were slightly lower but the difference was not significant, however they were significantly lower in the Baltic and Belt Sea. The results show correlation between arsenic concentrations and salinity for all studied species. Marine fish contain higher levels of arsenic than freshwater (Larsen and Francesconi 2003).

Polak-Juszczak (2013) analyzed the long-term trends for arsenic concentrations in liver tissues of Baltic cod in the period 1994–2010. The status of Baltic cod stock was monitored three times a year (winter-spring, summer, fall) in 30 locations within the Polish EEZ. Relatively high levels of arsenic in cod liver might reflect large quantities of this element transported with floods. Even though some annual differences occur, the mean values from 2010 (4.42 $\mu g\ g^{-1}$ dry weight) and 1994 (3.94 $\mu g\ g^{-1}$) do not differ significantly. To compare those results with previous studies, calculation for wet weight was made with approximate moisture content of cod liver about 68% (Niesteruk 1996) and the results are slightly higher than those obtained on cod muscle in previous studies. To conclude: no trends in increasing or decreasing of arsenic concentrations in time were observed in the case of cod. In studies from 2009 Polak-Juszczak observed a downward trend in arsenic concentrations in sprat tissues and no visible trend in herring samples.

Polak-Juszczak (2009), (2013) underlines that concentrations of arsenic in fish products originating from the Baltic Sea and available on Polish market are safe for consumers. Also the health risk associated with consumption of fish from locations close to the wrecks is assessed as low (Tørnes et al. 2002).

3 Conclusions

Although there are numerous scientific papers on arsenic and its speciation in marine environments and arsenic contamination appears as a global problem, not much research was conducted concerning this issue in the area of Baltic Sea.

Research concerning arsenic contamination in the Baltic Sea conducted within last few years was mainly related to inquisition of the chemical weapon dumped at the bottom after the World War II and referred only to the total arsenic measurements in sediments (Bełdowski et al. 2013, 2016a, ; Fauser et al. 2013; Khalikov and Savin 2011; Emelyanov et al. 2010; Garnaga et al. 2006; Garnaga and Stankevičius 2005) or were performed in the most contaminated areas of Bothnian Bay or Gulf of Finland (Vallius 2012, 2015; Yu et al. 2016). Collected data, however, did not refer to the whole Baltic area and there is only scarce data on the concentration of arsenic in sediments from e.g. Polish EEZ. Regarding marine organisms, the only research conducted nowadays in the Baltic Sea is related to fish.

Even though the concentrations of arsenic are relatively low in most compartments of the Baltic Sea environment, it is still worth observing and understanding the processes occurring in this ecosystem. Specificity of the Baltic Sea (steep salinity and temperature gradients, oxygen deficiency in deep basins and different sediment coverage) makes it impossible to compare the environment of Baltic Sea with oceanic ecosystems and the results of arsenic measurements described in literature concerning other regions. Also the issue of arsenic speciation was already raised by other researchers, studying both marine and freshwater ecosystems; however, so far there is no information on the behavior of arsenic species in marine waters with very low salinity—like the brackish waters of the Baltic Sea. Changing environment of the Baltic Sea may result in fluctuations of parameters controlling arsenic release from the sediments, including the most toxic arsenic species—arsenic (III). There are still some gaps in the knowledge on how arsenic behaves in the Baltic Sea, which could be filled by investigation of arsenic speciation and by spreading the study areas from contaminated sites only to the whole Baltic area.

Acknowledgements The article was compiled within a research grant "Arsenic speciation in the environment of southern Baltic Sea" nr 2016/21/N/ST10/03245 financed by the National Science Centre, Poland.

References

Andreae MO, Froelich PN (1984) Arsenic, antimony and germanium biogeochemistry in the Baltic Sea. Tellus series B—Chem Phys Meteorol 36:101–117

BACC (2015) Second assessment of climate change for the Baltic Sea Basin. Regional Climate Studies, Springer. doi:https://doi.org/10.1007/978-3-319-16006-1_1

Ballin U, Kruse R, Rüssel H-A (1994) Determination of total arsenic and speciation of arseno-betaine in marine fish by means of reaction headspace gas chromatography utilizing flame-ionizated detection and element specific spectrometric detection. Fresenius J Anal Chem 350:54–61

Bełdowski J, Szubska M, Emelyanov E (2013) Spatial variability of arsenic concentrations in Baltic Sea surface sediments in relation to sea dumped chemical munitions, E3S web of conferences 1, 16002. doi:https://doi.org/10.1051/e3sconf/20130116002

Bełdowski J, Szubska M, Emelyanov E, Garnaga G, Drzewińska A, Bełdowska M, Vanninen P, Östin A, Fabisiak J (2016a) Arsenic concentrations in Baltic Sea sediments close to chemical munitions dumpsites. Deep sea Res Part II: Top Stud Oceanogr 128:114–122

Bełdowski J, Klusek Z, Szubska M, Turja R, Bulczak AI, Rak D, Brenner M, Lang T, Kotwicki L, Grzelak K, Jakacki J, Fricke N, Östin A, Olsson U, Fabisiak J, Garnaga G, Rattfelt Nyholm J, Majewski P, Broeg K, Söderström M, Vanninen P, Popiel S, Nawała J, Lehtonen K, Belgrind R, Schmidt B (2016b) Chemical munitions search and assessment—an evaluation of the dumped munitions problem in the Baltic Sea. Deep Res Part II: Top Stud Oceanogr 128:85–95

Bissen M, Frimmel FH (2003) Arsenic—a review. Part I: occurrence, toxicity, speciation, mobility. Acta Hydrochim Hydrobiol 31(1):9–18

Blanck H, Holmgren K, Landner L, Norin H, Nottini M, Rosemain A, Sundelin B (1989) Advanced hazard assessment of arsenic in the Swedish environment in chemicals in the aquatic evironment. In: Lander L (ed), Springer

Borg H, Jonsson P (1996) Large-scale metal distribution in Baltic Sea sediments. Mar Pollut Bull 32:8–21

Cartensen K, Andersen JH, Gustafsson BG, Conley DJ (2014) Deoxygenation of the Baltic Sea during the last century. In: Proceedings of the national academy of science of the United States of America, vol 111, pp 5628–5633

CHEMSEA findings (2014) Available https://www.chemsea.eu/admin/uploaded/CHEMSEA%20Findings.pdf. Accessed 14 July 2017

Cullen WR, Reimer KJ (1989) Arsenic speciation in the environment. Chem Rev 89:713–764

Emelyanov EM (2007) The geochemical and geoecological situation of the Gotland Basin in the Baltic Sea where chemical munitionss were dumped. Geologija 60:10–26

Emelyanov E, Kravtsov V, Savin Y, Paka V, Khalikov I (2010) Influence of chemical weapons and warfare agents on the metal contents in sediments in the Bornholm Basin, the Baltic Sea. Baltica 23:77–90

Fausner P, Sanderson H, Hedegaard RV, Sloth JJ, Larsen MM, Krongaard T, Bossi R, Larsen JB (2013) Occurence and sorption properties of arsenicals in marine sediments. Environ Monit Assess 185:4679–4691

Flora (ed) (2015) Handbook on arsenic toxicology. Elsevier, Amsterdam. doi:https://doi.org/10.1016/B978-0-12-418688-0.00028-9

Francesconi KA (2010) Arsenic species in seafood: origin and human health implications. Pure Appl Chem 82:373–381

Francesconi KA, Schwerdtle T (2016) Fat-soluble arsenic—new lipids with a sting in their tail. Lipid Technol 28:96–98

Garnaga G, Stankevičius A (2005) Arsenic and other environmental parameters at chemical munitions dumpsite in the Lithuanian Economic Zone of the Baltic Sea. Environ Res Eng Manag 3:24–31

Garnaga G, Wyse E, Azemard S, Stankevičius A, de Mora S (2006) Arsenic in sediments from the southeastern Baltic Sea. Environ Pollut 144:855–861

Hall LW, Anderson RD (1995) The influence of salinity on the toxicity of various classes of chemicals to aquatic biota. Crit Rev Toxicol 25:281–346

Harkabusová V, Macharáčková B, Čelechovská O, Vitulová E (2009) Determination of arsenic in the rainbow trout muscle and rice samples. Czech J Food Sci 27:404–406

HELCOM (1995) Final report of the ad hoc working group on dumped chemical munitions (HELCOM CHEMU) to the 16th meeting of the Helsinki Commission (March 1995) I. HELCOM, p 20

HELCOM (2010) Hazardous substances in the Baltic Sea: an integrated thematic assessment f the hazardous substance in the Baltic Sea. In: Baltic Sea Environment Proceedings no 120B

Karbe L, Schnier CH, Siewers HO (1977) Trace elements in mussels (*Mytilus edulis*) from coastal areas of the North Sea and the Baltic sea. Multilement analyses using instrumental neuron activation analysis. J Radioanal Chem 37:927–943

Kasperek T (1997) Chemical weapons dumped in the Baltic Sea, ECE Toruń

Khalikov IS, Savin YI (2011) Arsenic content in water and bottom sediments in the areas of chemical weapon dumps in the Bornholm Basin of the Baltic Sea. Russ Meteorol Hydro+ vol 36:315–323

Knobloch T, Bełdowski J, Böttcher C, Söderström M, Rühl N, Sternheim J (2013) Chemical munitions dumped in the Baltic Sea. Report of the ad hoc expert group to update and review the existing information on dumped chemical munitionss in the Baltic Sea (HELCOM MUNI) Baltic Sea Environmental Proceedings. HELCOM, Baltic Sea Environment Proceedings 142:128

Kuenstl L, Griesel S, Prange A, Goessler W (2009) Arsenic speciation in bodily fluids of harbor seals (*Phoca vitulina*) and habor porpoises (*Phocoena phocoena*). Environ Chem 6:319–327

Larsen EH, Francesconi KA (2003) Arsenic concentrations correlate with salinity for fish taken from the North Sea and the Baltic waters. J Mar Biol Assoc UK 83:283–284

Lee B-G, Wallace WG, Luoma SN (1998) Uptake and loss kinetics of Cd, Cr and Zn in the bivalves *Potamocorbula amurensis* and *Macoma balthica*: effects of size and salinity. Mar Ecology prog Series 175:177–189

Ljunggren K, sjoestrand B, Johnels AG (1971) Activation analysis of mercury and other environmental pollutants in water and aquatic ecosystems. Nucl Techn Environ Pollut, 373–405

Maher W, Butler E (1988) Arsenic in the marine environment. Appl Organom Chem 2:191–214

McLusky DS, Bryant V, Campbell R (1986) The effect of temperature and salinity on the toxicity of heavy metals to marine and estuarine invertebrates. Oceanogr Mar Biol Annu Rev 24: 481–520

MERCW (2006) Modelling of ecological risk related to sea-dumped chemical weapons—synthesis paper on available data. Available http://www.mercw.org/images/stories/pdf/synthesis_d2.1.pdf. Accessed 14 July 2017

Missiaen T, Söderström M, Popescu I, Vanninen P (2010) Evaluation of a chemical munitions dumpsite in the Baltic Sea based on geophysical and chemical investigations. Sci Total Environ 408:3536–3553

Moreira A, Freitas R, figueira E, Volpi Ghirardini A, Soares AMVM, Radaelli M, Guida M, Libralato G (2018) Combined effects of arsenic, salinity and temperature on *Crassostrea gigas* embryotoxicity. Ecotoxicol Environ Saf 147:251–259

Nfon E, Cousins IT, Jorvinen O, Mukherjee AB, Verta M, Broman D (2009) Trophodynamics of mercury and other trace elements in a pelagic food chain from the Baltic Sea. Sci Total Environ 407:6267–6274

Nicholas DR, Ramamoorthy S, Palace V, Spring S, Moore JN, Roseznzweig RF (2003) Biogeochemical transformations of arsenic in circumneutral freshwater sediments. biodegradation 14:123–137

Niedzielski P, Siepak M, Siepak J (2000) Występowanie i zawartość arsenu, antymonu i selenu w wodach i innych elementach środowiska. Rocznik Ochrona Środowiska tom 2, pp 317–341; Środkowo-Pomorskiej Towarzystwo Naukowe Ochrony środowiska (in polish)

Niesteruk R (1996) Właściwości termofizyczne żywności. Rozprawy naukowe Politechnika Białostocka vol 33 (in polish)

Paka V, Spridonov M (2001) Research of dumped chemical weapons made by R/V "Professor Stockman" in the Gotland, Bornholm and Skagerrak Dumpsites. In: Missiaen T (ed) Chemical munitions dump sites in the coastal environments, pp 27–42

Polak-Juszczak L (2009) Temporal trends in the bioaccumulation of trace metals in herring, sprat and cod from the southern Baltic Sea in the 1994–2003 period. Chemosphere 76:1334–1339

Polak-Juszczak L (2013) Trace elements in the livers of cod (*Gadus morhua* L.) from the Baltic Sea: levels and temporal trends. Environ Monit Assess 185:687–694

Preis S (1997) Environmental problems of the Baltic Sea region. In: Glantz MH, Zonn IS (eds) NATO ASI series: Scietific, Environmenal and political issues in the circum-Caspian Region, vol 29 pp 227–244

Rahman MA, Hasegawa H, Lim RP (2012) Bioacummulation, biotransformation and trophic transfer of arsenic in the aquatic food chain. Environ Res 116:118–135

Reimann C, Matschullat J, Birke M, Salminen R (2009) Arsenic distribution in the environment: The effect of scale. Appl Geochem 24:1147–1167

Salminen R (ed) (2005) Geochemical atlas of Europe. Available http://weppi.gtk.fi/publ/foregsatlas/index.php. Visited on 15 July 2017

Sanderson H, Fauser P, Thomsen M, Sørensen PB (2008) Screening level fish community risk assessment of chemical warfare agents in the Baltic Sea. J Hazard Mater 154:846–857

Sioen I, De Henauw S, Van Camp J, Volatier J-L, Leblanc J-C (2009) Comparison of the nutritional-toxicological conflict related to seafood consumption in different regions worldwide. Regul Toxicol Pharmacol 55:219–228

Snoeijs-Leijonmalm P, Schubert H, Radziejewska T (2017) Biological oceanography of the Baltic Sea, Springer. doi:https://doi.org/10.1007/978-94-007-0668-2_1

Stoeppler M, Burow M, Backhaus F, Schramm W, Nürnberg HW (1986) Arsenic in seawater and brown algae of the Baltic Sea and the North Sea. Mar Chem 18:321–334

Szczepańska T, Uścinowicz S (1994) Geochemical atlas of the Southern Baltic Sea. Warszawa 1994 (in polish)

Taylor V, Goodale B, Raab A, Shwerdtle T, Reimer K, Conklin S, Karagas MR, Francesconi KA (2017) Human exposure to organic arsenic species from seafood. Sci Total Environ 580: 266–282

Tornero V, Hanke G (2016) Chemical contaminants entering the marine environment from sea-based sources: a review with a focus on European seas. Mar Pollut Bullet 112:17–38

Tørnes JA, Voie ØA, Ljønes M, Opstad AM, Bjerkeseth LH, Hussain F (2002) Investigation and risk assessment of ships loaded with chemical ammunition scuttled in Skagerrak. TA-1907/2002

Turja R, Hoher N, Snoeijs P, Barsiene J, Butrimaviciene L, Kuznetsova T, Kholodkevich SV, Devier M-H, Budzinski H, Lehtonen KK (2014) A multibiomarker approach to the assessment of pollution impacts in the Baltic Sea coastal areas in Sweden using caged mussels (*M. trossulus*). Sci Total Environ 473–474:398–409

Turja R, Lehtonen KK, Meierjohann A, Brozinksi J-M, Vahtera E, Soirinsuo A, sokolova, Snoeijs P, Budzinski H, Devier M-H, Peluhet L, Pääkkönen J-P, Viitasalo M, Kronberg L (2015) The mussel caging approach in assessing biological effects of wastewater treatment plant discharges in the Gulf of Finland (Baltic Sea). Mar Pollut Bull 97:135–149

Us EPA (2010) Study of discharges incidental to normal operation of commercial fishing vessels and other non-recreational vessels less than 79 ft. Report to Congress, Washington, D.C.

Uścinowicz S (2011) Geochemistry of Baltic Sea surface sediments. Polish Geological Institute

Usydus Z, Szlinder-Richert J, Polak-Juszczak L, Komar K, Adamczyk M, Malesa-Ciecwierz M, Ruczyńska W (2009) Fish products available in Polish market—assessment of nutritive value and human exposure to dioxins and other contaminants. Chemosphere 74:1420–1428

Vallius H (1997) The distribution of some heavy metals and arsenic in recent sediments from the eastern Gulf of Finland. Appl Geochem 13:369–377

Vallius H (2012) Arsenic and heavy metal distribution in the bottom sediments of the Gulf of Finland through the last decades. Baltica 25:23–32

Vallius H (2014) Heavy metal concentrations in sediment cores from the northern Baltic Sea: Declines in the last two decades. Mar Pollut Bull 79:359–364

Vallius H (2015) Applying sediment quality guidelines on soft sediments of the Gulf of Finland, Baltic Sea. Mar Pollut Bull 98:314–319

Vallius H, Lehto O (1998) The distribution of some heavy metals and arsenic in recent sediments from the eastern Gulf of Finland. Appl Geochem 13:369–377

Vallius H, Leivuori M (1999) The distribution of heavy metals and arsenic in recent sediments in the Gulf of Finland. Boreal Environ Res 4:19–29

Vallius H, Ryabchuk D, Kotilainen A (2007) Distribution of heavy metals and arsenic in soft surface sediments of the coastal area off Kotka, northeastern Gulf of Finland, Baltic Sea. Geol Surv Finl Spec Pap 45:33–48

Widerlund A, Ingri J (1995) Early diagenesis of arsenic in sediments of the Kalix River estuary, northern Sweden. Chem Geol 125:185–196

Yli-Hemminki P, Sara-Aho T, Jørgensen K, Lehtoranta J (2016) Iron-manganese concretions contribute to benthic release of phosphorus and arsenic in anoxic conditions in the Baltic Sea. J Soils Sedim 16:2138–2152

Yu C, Peltola P, Nystrand MI, Virtasalo JJ, Österholm P, Ojala AEK, Hogmalm JK, Åström ME (2016) Arsenic removal from contaminated brackish sea water by sorption onto Al hydroxides and Fe phases mobilized by land-use. Sci Total Environ 542(A):923–934

Zaborska A, Kosakowska A, Bełdowski J, Bełdowska M, Szubska M, Walkusz-Miotk J, Żak A, Ciechanowicz A, Wdowiak M (2016) The distribution of heavy metals and ^{137}Cs in the central part of the Polish maritime zone (Baltic Sea)—the area selected for wind farm acquisition. Estuarine, Coastal and Shelf Science 198:471–481 https://doi.org/10.1016/j.ecss.2016.12.007

Protists of Arctic Sea Ice

Z. T. Smoła, A. M. Kubiszyn, M. Różańska, A. Tatarek and J. M. Wiktor

Abstract Sea ice not only shapes the global climate but is also an important background for a complicated ecosystem that is closely related to the littoral benthic ecosystem. This similarity is the reason why this formation is usually referred to as an "inverted bottom." In the deep central part of the Arctic Basin (which is 47% of its overall surface area), it is estimated that approximately 50% of the primary production comes from autotrophic protists (sympagic) related to sea ice. Global warming has caused changes in the range and time of sea ice occurrence, and the existence time of sea ice assemblages is also changing. After 173 years of ice-related microalgae studies, the appearance of 1027 taxa closely related to sea ice has been recorded.

Keywords Protists · Arctic · Sea ice · Primary production

Z. T. Smoła (✉) · A. M. Kubiszyn · M. Różańska · A. Tatarek · J. M. Wiktor
Marine Ecology Department, Institute of Oceanology Polish Academy of Sciences,
ui. Powstańców Warszawy 55, 81–712 Sopot, Poland
e-mail: zosiasmola@iopan.gda.pl; zosiasmola@gmail.com

A. M. Kubiszyn
e-mail: derianna@iopan.gda.pl

M. Różańska
e-mail: rozanskam@yahoo.com

A. Tatarek
e-mail: aniak@iopan.gda.pl

J. M. Wiktor
e-mail: wiktor@iopan.gda.pl

Z. T. Smoła
Centre for Polar Studies KNOW (Leading National Research Centre),
Faculty of Earth Sciences, University of Silesia,
ui. Będzińska 60, 41-200 Sosnowiec, Poland

T. Zielinski et al. (eds.), *Interdisciplinary Approaches for Sustainable Development Goals*, GeoPlanet: Earth and Planetary Sciences,
https://doi.org/10.1007/978-3-319-71788-3_10

1 Introduction

During periods of maximum reach, sea ice covers approximately 12% of the global ocean surface (Weeks 2010). The extensive structure of sea ice creates a type of "inverted bottom" system that is well suited for colonization by protists. The protists in the water column continuously grapple with gravity, the viscosity of the medium and the mixing of the water column. Substrate stability provided by sea ice is extremely important for autotrophic organisms, as they can develop in the sea ice without the risk of falling below the overexposed layer of the sea (the euphotic zone). On the other hand, ice cover acts as a "curtain" that weakens the transmission of light to the under-ice layers. Low values of photosynthetically active radiation (PAR) are compensated by relatively constant levels, which does not hinder the development of autotrophic protists adapted to low levels of PAR. Autotrophic microalgae can settle on the surface and within the porous structure formed by saltwater channels, which are characteristic of sea ice. In contrast to the seabed, movement by water currents or wind-driven ice serves as a vector for transferring coastal residents to the almost unproductive waters of the deep central Arctic Basin.

2 Sea Ice

The Arctic, unlike the Antarctic, is an environment dominated by the sea. The Arctic, surrounded by land is called by some Arctic Mediterranean Sea. The total area of the Arctic Ocean exceeds 14 million km^2 and represents approximately 4% of the global ocean surface. From the autumn to the late spring, the Arctic Sea is mostly covered with sea ice, and the maximum reach occurs in March.

In the age of progressing global warming, sea ice cover has decreased from decade to decade (Kwok and Rothrock 2009). At the end of 2012, sea ice only had an area of 3.6 million km^2 (Fig. 1). The area decreased by 36% from 2000, which represented approximately 5% of the northern hemisphere (Maykut 1985). Ice cover is composed of 80–85% of the sea ice formed at the end of the astronomical winter. The remainder is perennial ice that has survived the melting period, which usually starts in May.

Qualitative changes are simultaneously observed with the loss of sea ice. In the 1980s, 40% of ice was composed of perennial ice, while it has dropped to just 10% in the current decade (Fig. 2). Pessimistic prognoses assume that the perennial ice in the Arctic Basin will completely disappear by 2050 (Assessment 2009). The optimists, however, claim that this will occur at the end of the current century (Kwok et al. 2009).

To understand the role of sea ice in the life of the communities associated with it, it is necessary to take a closer look at the processes of its formation. Sea ice begins to form only when sea water (at 35 PSU) cools to approximately −1.8 °C and to−1.3 °C at approximately 24 PSU. The density of sea water that is cooled at the surface in contact with air will increase and gradually fall to the deeper parts of the water column. That dense and cold water is replaced by warmer and lighter water masses at the

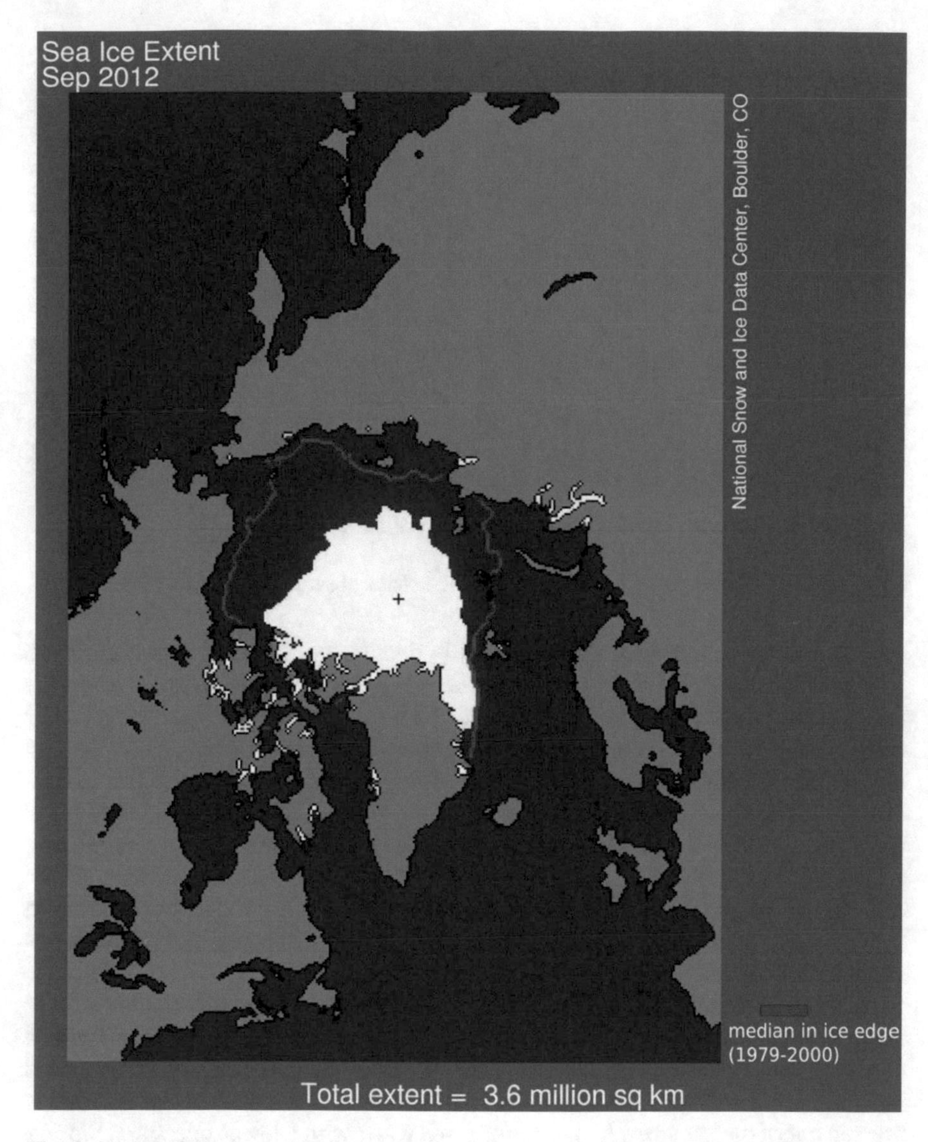

Fig. 1 Extent of Arctic sea ice. *Source* National snow and ice data center

surface, which also fall to deep layers after cooling. This process continues until the water in the top layer of the water column (50–100 m) cools to the freezing point. First, tiny ice needles (frazil ice) are formed, which are lighter than the surrounding sea water. They flow to the surface and form a delicate, "oily", grease ice coating in windless weather, which thickens as it freezes. During windy weather, pancake ice is formed, consisting of more or less round, disc shaped ice pieces with diameters of 0.2–3 m. When floating to the surface, ice needles act as a plankton net, "scraping" suspended particles in the water. Some authors, e.g., Parker et al. (1985), believe that

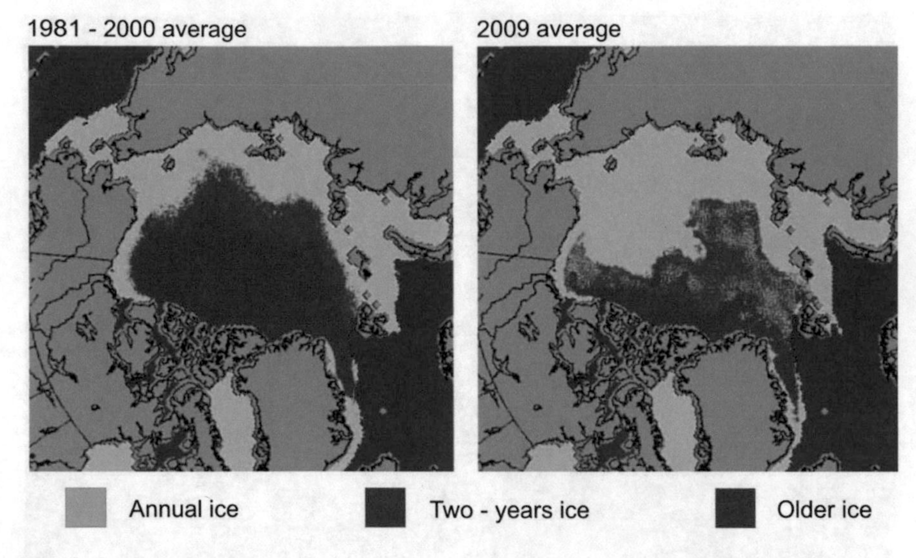

Fig. 2 Qualitative changes to Arctic sea ice. *Source* National snow and ice data center

the suspension—along with the protist cells floating in the water column, acts as condensation nuclei, by which cooled sea water begins to freeze. Both sea ice forms (grease and pancake) combine into a larger compact coating that isolates the water column from the air and causes the ice to continue to grow only from below, thereby preserving the previously seized particles. After a few days, young ice (more than 10–30 cm thick) is formed. The ice development process lasts from late fall to May/June. As a result of the insulating properties of ice and the snow layer accumulating on its surface, the rate of increase in the sea ice thickness declines over time. By the end of the spring in the Arctic (April/May), the average thickness of sea ice is up to 2 m (at most up to 3 m), and in warmer areas (e.g., West Spitsbergen), its thickness exceeds one metre.

The ice that has formed appears to be homogeneous at the surface and is similar to a pane of glass. However, it is not. During the formation of sea ice, water forms a crystal structure that does not leave any space for additives such as salt. This means that the salts dissolved in the water must be removed outside the ice. The salt that is pressed out of the ice forms higher salinity sea water with a lower freezing point. As a result, the apparently homogeneous sea ice is, in fact, a heterogeneous structure. Its porosity, which is calculated by the total surface of the unfrozen channels filled with brine, is from 0.6 to 4 $m^2\ kg^{-1}$ (Krembs et al. 2000). The salinity in brine channels may reach more than 100 PSU (Spindler 1996; Bartsch 1989). Values of the 40 PSU (Schünemann 2004), thus exceeding the salinity of the Dead Sea, are not extraordinary. Despite the high salt concentrations in brine, the general salinity in the core of one-year sea ice is relatively low, ranging from 2 to 14 PSU (Maykut 1985), and the salinity is lower if the ice is older. This results in a strong stratification of the water column under ice, with a marked layer of low salinity water at the time of melting. The changes in the salt concentrations associated with the formation and melting of sea ice cause ice-dwelling organisms to face extremely high salinities in the autumn and low salinities in the spring and summer.

2.1 *Factors Affecting the Development of Sea Ice Protists*

2.1.1 Light

Ice cover is a surface that constitutes an effective barrier for the sun's rays. In circumpolar regions where the supply of light is limited even without the effects of ice, it is a factor that significantly affects the life under the sea. In the Arctic region, especially at high latitudes, extreme light seasonality occurs. This seasonality is noticeable above 60° latitude, where an imbalance between the lengths of day and night is clearly visible (from 6 "bright" hours in the winter to 19 in the summer), and the sun does not rise more than 53° above the horizon. Above the Arctic Circle, the length of the bright period varies from 100 to 200 days. For the remainder of the year, it is at least twilight except for a brief period of days and nights. For this reason, the average annual PAR dose is only 23% of the dose that reaches the surface of the equatorial region.

The low position of the sun over the horizon is, in other words, the acute angle of the sun rays, resulting in a strong reflection of light (albedo) from the surface of the sea, which is multiplied by the presence of ice and snow. The undisturbed sea surface reflects 4% of incident radiation at 40° and approximately 42% at 5° incidence (Jerlov 1968). Despite the shrinking sea ice cover in the Arctic, the covered area is still significant. Thus, ice reflects 15 to 90% of the incident radiation. The remaining non-reflected part of the radiation is further weakened as it passes through the snow layer and the ice column. In such a system, the snow dominates the weakening of the passing light. The value of the attenuation coefficient varies depending on the type of the snow (according to the Inuit—the indigenous people of the Arctic and Subarctic areas—there are more than 100 types). The attenuation coefficient varies from 1.5 to 45 m^{-1} (Palmisano et al. 1987), and the coefficient for the ice column does not exceed 1.6 m^{-1}. Depending on the topography of the snow-ice cover, 0.15–7% of the PAR reaches the surface of the sea. The significant total light attenuation has a clear effect on the condition and abundance of sea ice protist (Fig. 3).

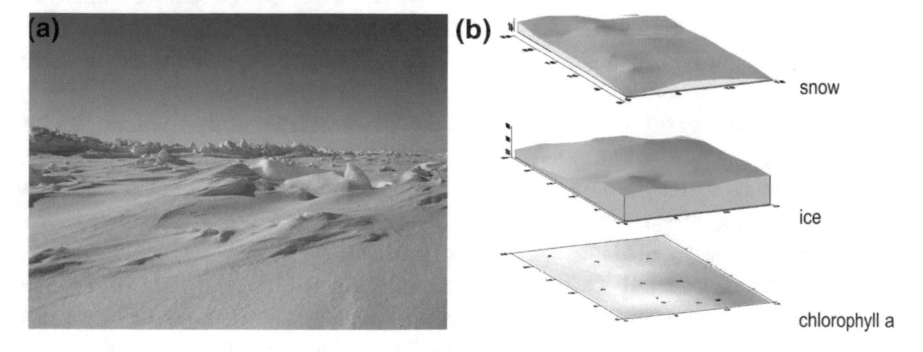

Fig. 3 Differentiation of snow-ice cover thickness and its effect on the autotrophic protists associated with ice. **a** View of a consolidated ice pack. *Photo* J. Wiktor **b** Measurement based on data from the CASES project (Patchiness experiment)

2.1.2 Hydrological Factors

Other important factors affecting protist development in sea ice are nutrient availability and metabolic activity in low temperatures. Although carbon fixation and the growth of ice protist communities are most efficient in temperatures ranging from 4 to 14 °C (Thomas and Dieckmann 2000), these organisms develop the ability to survive in environments where the temperature may reach −16 °C. Sea ice protists do not freeze internally because of their ability to accumulate a special kind of proteins that prevent ice crystallization—antifreeze proteins (AFPs). It is assumed that the most stable and favourable physical and chemical conditions for protist development are in the lower—approximately 20 cm—layer of the ice, where the most abundant protist assemblages associated with sea ice dwell.

3 Protists

In 1841, Ehrenberg (1841) initiated studies on ice protists and published a list of diatom species that he observed in the sea ice in the area of Spitsbergen and the Barrow Strait in 1853 (Ehrenberg 1853). Therefore, protists have been studied for over 170 years. Three-quarters of a century later, Bursa (1961) described the ice diatom assemblages of Hudson Bay (Canada). The first studies were executed from ships frozen in sea ice, often for more than one year (sometimes forever). Since then, as the technology has advanced, studies on ice protists have increased. Currently, surveys are based from icebreakers, airplanes and helicopters landing on ice, and progress in materials technology has facilitated the participation in field work.

Despite such a long tradition of research, the question “where do microalgae come from in sea ice” remains, since ice is formed during the polar night or at the end of the polar day period.

To understand this, the “sea ice” section must be revisited, where the processes related to sea ice formation are described. As mentioned earlier, the initial stage of ice formation begins in the fall when few representatives of microalgae remain in the water as residue from the previous growing season. Autotrophs are inactive due to the lack of light but remain as living cells in the water. Tiny needles of frazil ice float to the surface and grab particles from the water column (Garrison 1991), including bacteria, various nano- (smaller than 20 μm) and microplanktonic (larger than 20 μm), photo (autotrophic) and phagotrophic (heterotrophic) organisms. Their abundance usually is low, on the order of one thousand per square metre. In newly formed ice, the abundance of protists can be fifty times higher than in the surrounding sea water (Thomas and Dieckmann 2000).

The migration of frazil ice towards the surface is not the only reason for the concentration of protist in sea ice. Cells can also be caught even after ice cover has formed, which occurs via the comb-like structure of the lowermost layer of the sea

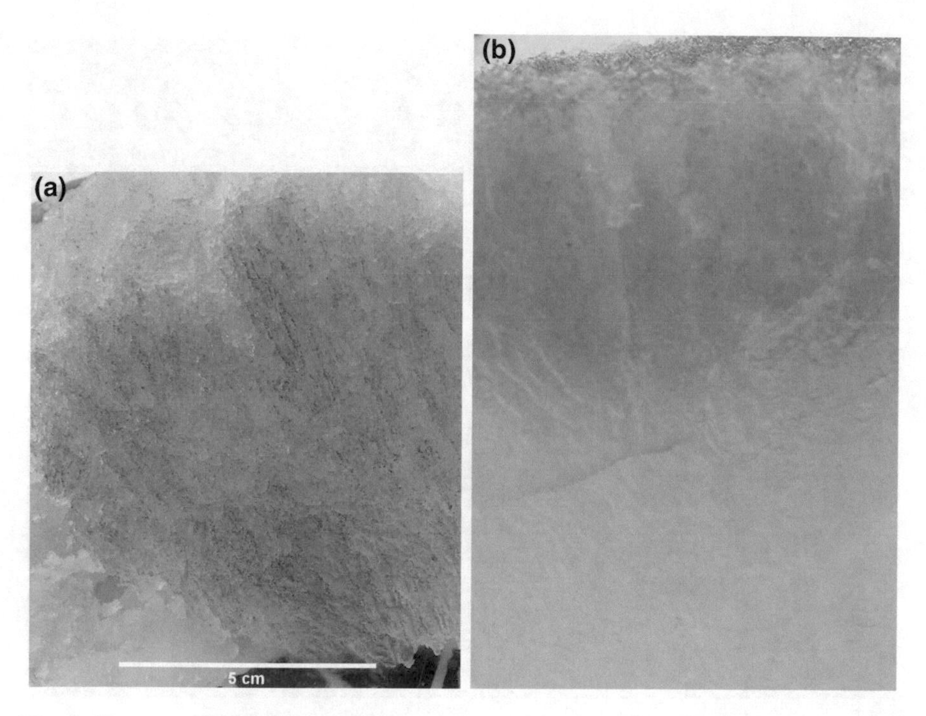

Fig. 4 Structures of the lower layer of the sea ice (**a**) and brine channels (**b**). *Photos* J. Wiktor

ice. In addition, due to waves and sea tides, the water pressure pushes the microalgae cells into the open saline channels (Fig. 4a, b). So far, studies have demonstrated that the incorporation of cells into the ice during formation is selective, and not all cells survive the drastic changes in the environment (Fig. 5).

Significant developments in sea ice studies have finally allowed the first panarctic list of ice taxa to be generated. So far, 1027 taxa of protists have been indicated,

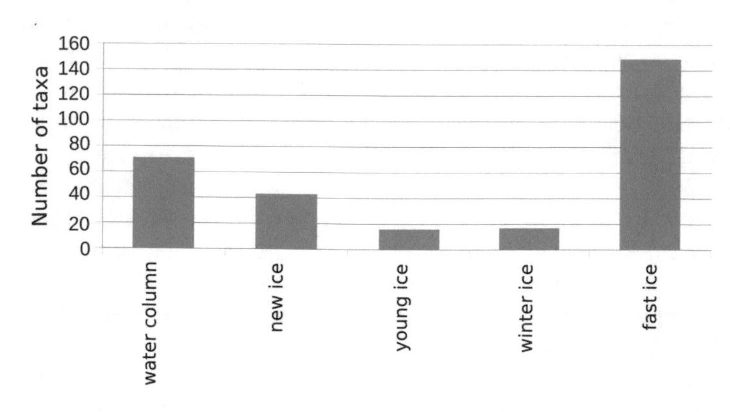

Fig. 5 Cell concentration in the formed sea ice of Franklin Bay (based on data from Rozanska et al. 2009)

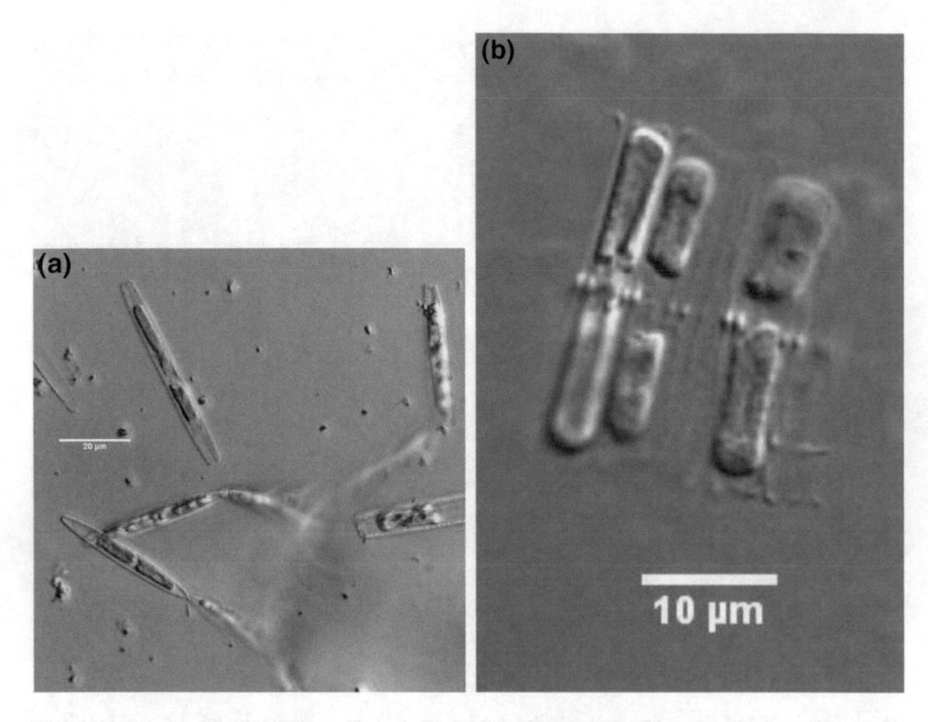

Fig. 6 Taxa associated with Arctic sea ice—*Nitzschia frigida* (**a**), *N. pelagica* (**b**). *Photos* J. Wiktor

among which diatoms are the most dominant group (71%) (Poulin et al. 2011), which are most commonly represented by Pennales. Criopelagic taxa are present in ice assemblages, such as shrubby colonies of *Nitzschia frigida* (Fig. 6a) that are associated with ice and are capable of living in the pelagic environment, band-like colonies of *Navicula pelagica* (Fig. 6b), *Fragilariopsis cylindrus*, and *F. oceanica. Pauliella taeniata* has also been recorded during the spring in the Baltic Sea. Moreover, "heavier", poorly floating representatives belonging to benthic genera such as Pleurosigma (Fig. 7a) and *Navicula transitans* have also been found in brine channels. Melosiraceae *M. arctica* (Fig. 7b) has also been associated with ice, which in the summer forms large, branched garlands of algae mats under the sea ice that can reach 3 m in length (Melnikov 1997). Diatoms inhabiting the sea ice are usually relatively large (more than 20 μm); hence, single cells, as well as colonies, are important food bases for the higher trophic levels. The most common species of the centric (Centrales) diatoms is *Attheya septentrionalis*; however, because of the small size (even for protist), these diatoms rarely dominate the biomass of ice communities. Other representatives of Centrales are mainly taxa belonging to the genus *Thalassiosira* sp. In addition, *Chaetoceros* sp. occur only in young ice.

In contrast to pelagic areas, where the biomass and proportion of higher taxonomic groups of autotrophic protists can be roughly estimated from satellite observations (Comiso 2010), researchers must rely on field work to determine the

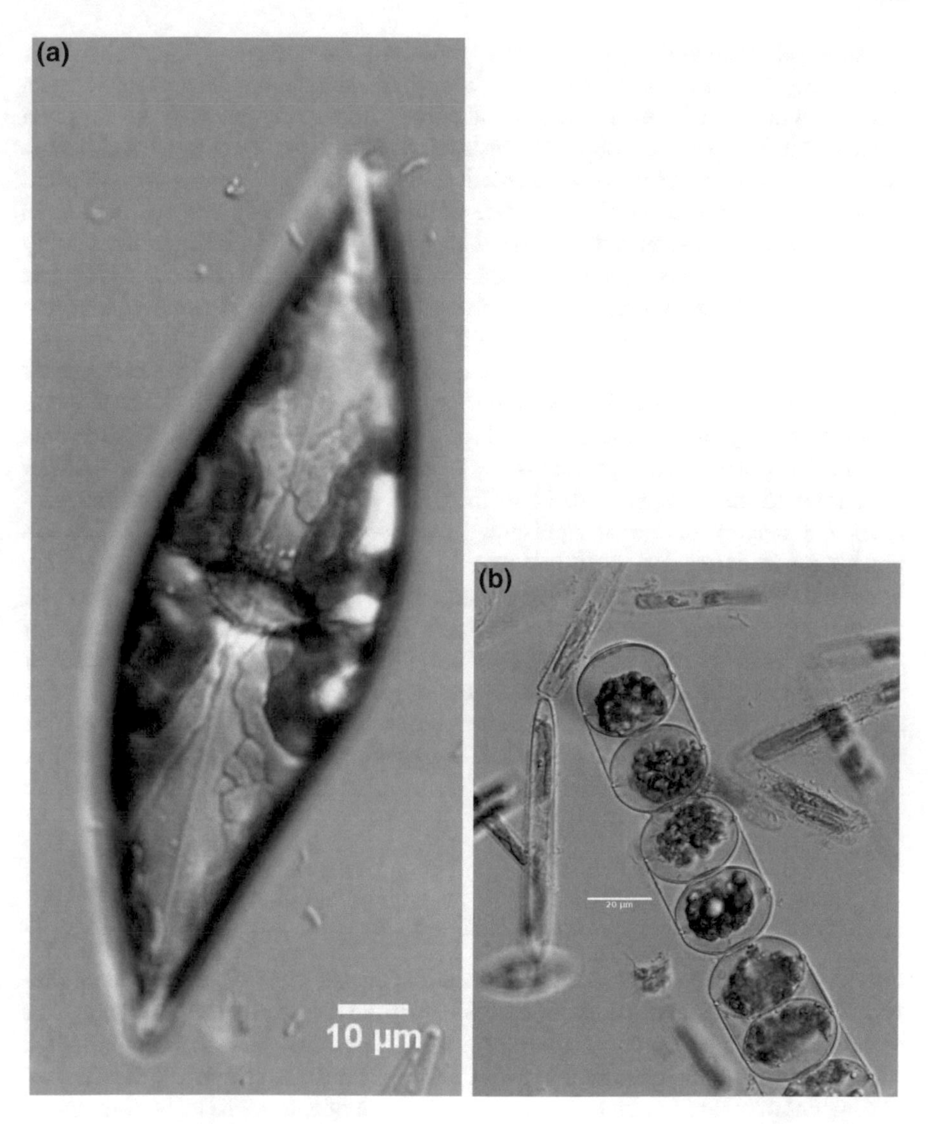

Fig. 7 Taxa associated with ice—*Pleurosigma* sp. (**a**), *Melosira arctica* (**b**). *Photos* J. Wiktor

composition of ice algae. It is not possible to take samples without time-consuming and labourious ice cores collection, which is so burdensome that the number of samples is usually limited. The most common tools used to sample ice are cores. This method has advantages as well as disadvantages. Sampling cores sometimes cause damage to the lower, brittle, inhabited layer of the ice, and the quality of the sample depends on the operator's precision. Sea ice sampling can also be conducted by a diver. This is a more onerous method (especially for the diver) and has greater logistical challenges, especially because during the most intensive development of the surveyed communities, the air temperature does not exceed—1 °C. The diver takes samples using cores or pumps, which are similar to underwater vacuum cleaners with a plankton net instead of a bag. Then, the collected ice must be melted. As the sea ice column is characterized by low salinity, considerable amounts of low saline water is released during thawing, exposing the collected organisms to osmotic shock. To prevent this, ice core samples are melted in pre-filtered (to remove planktonic organisms) marine water to maintain the salinity typical for the natural environment. Then, the prepared samples are ready for further analysis. The whole ice sample preparation process is often very invasive and often affects the quality of the results.

Figure 3 illustrates how variable the ice and snow cover are, and how they result in a patchy distribution of ice communities, which causes uncertainty in the estimations of the protist abundance and biomass. The distribution of sea ice protist communities also depends on the history of the formation of the sea ice, consumer pressure, and the movement of the drifting ice with the currents, which differentiates the density of the assemblages. It causes unpredictable results that are impossible to extrapolate from one area to another—even to the nearest region. Despite the difficulties in the accurate estimation of the abundance of ice protists, we are able to determine that the numbers fluctuate by at least five orders of magnitude ($<10^4$ to $>10^9$ under one square metre of ice) (Fig. 8). In general, lower values are characteristic of young ice or the upper layer, while high values refer to the ice-water contact layer and the internal spaces in ice that are filled with marine water (Thomas and Dieckmann 2000). A high abundance of protists is often visible from the colour of the sea ice (Fig. 4).

3.1 Primary Production

Protist biomass is estimated during microscopic analysis or from measured chlorophyll *a* concentrations with the assumption that most of the unicellular organisms associated with the sea ice are autotrophs. Protist biomass expressed in chlorophyll *a* concentrations has ranged from 0 to 300 mg m^{-2} in the Arctic. While considering the amount of organic carbon, the annual production of autotrophs inhabiting sea ice ranged from 6 to 73 TgC $year^{-1}$ (Thomas and Dieckmann 2000), and constitutes from 1 to 12.5% of the total primary production in the Arctic (data refer to pelagic primary production estimated by Arrigo et al. 2008). Although compared to pelagic organisms, the amount of organic carbon assimilated by autotrophic ice protists is relatively small. In areas covered with sea ice, they are the exclusive source of organic carbon available for higher trophic levels (Thomas and Dieckmann 2000). Although diatoms are the most abundant group in sea ice assemblages and have major contributions to the primary production, the significant role of other organisms in the functioning of sea ice communities has been discovered, including small autotrophic flagellates such as *Prymnesiophyceae*, *Dinoflagellata*, *Chlorophyta* (particularly *Prasinophyceae*), *Chrysophyceae* and *Cryptophyceae*.

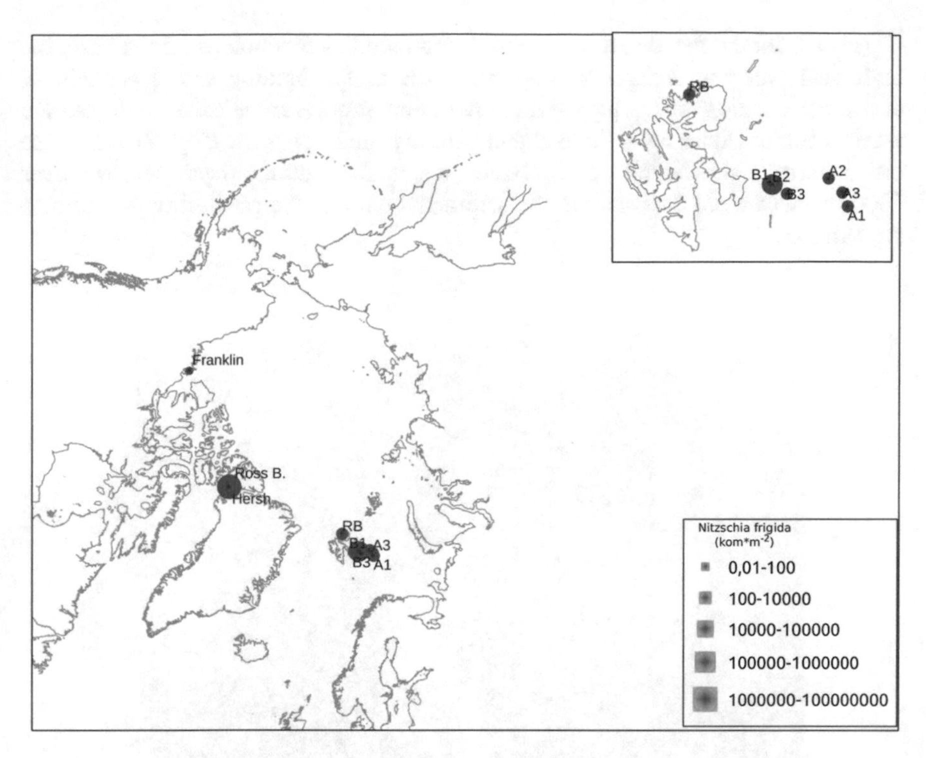

Fig. 8 Distribution and abundance of *Nitzschia frigida* in different Arctic regions (Wiktor 2015)

3.2 Beneficiaries of Primary Production

Recently, more attention has been paid to the role of the microbiological loop in the function of communities associated with sea ice. Phagotrophic protists (feeding on previously produced organic matter) control the abundance of bacteria and other unicellular organisms. Through the decomposition of organic matter, simple inorganic compounds are recycled in the environment and become available again for autotrophs. The most important contributors to the total biomass of phagotrophs are flagellates (cells with one or more flagella), including euglenids, *dinoflagellates*, ciliates and tiny nanoflagellates whose taxonomic affiliations have not been determined. Phagotrophic flagellates are a very diverse group of organisms that includes six distinct 'supergroups' of protists, of which the best known are *dinoflagellates* because of their significant abundance in the environment, considerable size and characteristic morphological features. In the sea ice, *dinoflagellates* are most often represented by the species of the *Protoperidinium* sp., *Gymnodinium* sp., *Gyrodinium* sp. and *Polykrikos* sp. genera. These species constitute an important food base for the multicellular organisms associated with the sea ice—from tiny crustaceans (including *Calanoida*, *Amphipoda*) to fish feeding on them (e.g., polar cod *Boreogadus saida*).

Apart from the fact that ice-associated communities function as a food base, the biological pump also plays an important role in the binding and deposition of excess carbon dioxide. During the melting period, protists are released to the marine water column. Under light, thermal and osmotic stress, cells die (Fig. 9) and fall to the bottom (e.g., Central Arctic Basin where the bottom depth reaches over 2000 m), where are deposited in the sediments without the possibility of return to the surface.

Fig. 9 Aggregates of dead sea ice protists during the ice melting period. *Photo* J. E. Ross

4 What Is Next?

In the time of shrinking sea ice cover, we have to face the question of what comes next for the sea ice associated protists. It could be said that since the ice will disappear, there will also be fewer protists. However, the answer to this question may not be so simple. The answer is complicated because the most severe loss of sea ice has been recorded in the summer. It is often said that the record (usually the minimum) extent of sea ice is recorded at the end of the season in September when the ice begins to form again. Sea ice formation, which is often late, continues in the spring during the time of autotroph blooms. Therefore, protists will not lose too much of their "favourite" habitat.

However, the substrate itself is not everything. The analysis of ice sheet anomalies indicates a decrease compared to the 1979–2000 average (NSIDC). Ice decline is greatest in the most productive Arctic shelf areas. Lower compactness of the ice cover results in a higher total surface area of the crevasses (windows). Although sea ice associated autotrophic organisms may have more light available, excess light can destroy shade-loving organisms. Dead diatom cell aggregates have been observed near crevasses by the authors. An example of how light adversely affects the abundance of autotrophs is presented as the temporary variability of protist obtained during studies in northern Svalbard (Fig. 10). Another scenario of climate change predicts the appearance of a thicker layer of snow on thinner ice; thus, the amount of light reaching the lower layer of the sea ice will decrease.

Acknowledgement The team of authors has participated in research on ice protists in international programmes such as NOW (North Water Polynya—1998), 'Marinok' (Project on Ice Margin Zone on Barents Sea)—1999–2000, CASES (Canadian Arctic Shelf Exchange Study)—2003/4, CLEOPATRA (Climate effects on planktonic food quality and trophic transfer in Arctic Marginal Ice Zones) 2007/2008, Resolute 2010/11 and have participated in ArcticNet (grant NR695/N-ARCTICNET/2010/0).

This publication was financed by funds from the Leading National Research Centre (KNOW) received by the Centre for Polar Studies for the period 2014–2018.

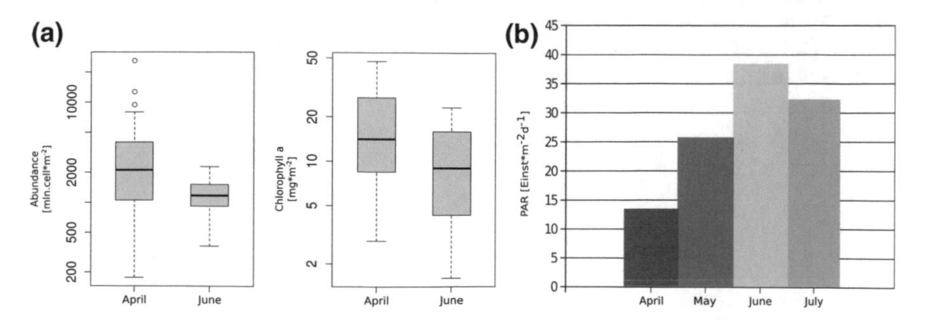

Fig. 10 Temporal variation the number of cells in sea ice associated assemblages (**a**) and PAR doses in Ripfjorden (Nordaustlandet) (**b**) (*Source* Data based on database of Cleopatra project)

References

Arrigo KR, Dijken G, Pabi S (2008) Impact of a shrinking Arctic ice cover on marine primary production. Geophys Res Lett 35(19):L19603

Assessment, Arctic marine shipping (2009) *Arctic Marine Shipping Assessment 2009 Report*, Arctic Council

Bartsch A (1989) Sea ice algae of the Weddell Sea (Antarctica): species composition, biomass, and ecophysiology of selected species. Ber Polarforsch 63:1–110

Bursa AS (1961) The Annual oceanographic cycle at Igloolik in the Canadian Arctic: II The Phytoplankton. J Fish Board Can 18(4):563–615

Comiso J (2010) Satellite remote sensing techniques, Polar Oceans from Space, pp 73–111

Ehrenberg CG (1841) 1853 Einen Nachtrag zu dem Vortrage uber Verbreitung und Einfluss des microscopischen Lebens in Sud-und Nord-Amerika, Acad. Wiss. Berlin Monatsber, pp. 220

Garrison DL (1991) An overview of the abundance and role of protozooplankton in Antarctic waters. J Mar Syst 2(3):317–331

Jerlov NG (1968) Optical oceanography. Elsevier Pub. Co, Amsterdam

Krembs C, Gradinger R, Spindler M (2000) Implications of brine channel geometry and surface area for the interaction of sympagic organisms in Arctic sea ice 243(1):55–80

Kwok R, Rothrock DA (2009) Decline in Arctic sea ice thickness from submarine and ICESat records: 1958–2008. Geophy Res Lett 36

Kwok R, Cunningham GF, Wensnahan M, Rigor I, Zwally HJ, Yi D (2009) Thinning and volume loss of the Arctic Ocean sea ice cover: 2003–2008. J Geophys Res 114(C7):C07

Maykut GA (1985) The Ice Environment. CRC Press Inc., Boca Raton, pp 21–82

Melnikov IA (1997) The Arctic sea ice ecosystem. Antarctic Sci 9(4):457–458

Palmisano A, SooHoo J, Moe R, Sullivan C (1987) Sea ice microbial communities. VII Changes in under-ice spectral irradiance during the development of Antarctic sea ice microalgal communities. Mar Ecol Prog Series 35:165–173

Parker LV, Sullivan CW, Forest TW, Ackley SF (1985) Ice nucleation activity of antarctic marine microorganisms. Antarctic J 20:126–127

Poulin M, Daugbjerg N, Gradinger R (2011) The pan-Arctic biodiversity of marine pelagic and sea-ice unicellular eukaryotes: a first-attempt assessment. Mar Biodivers 41:13–28

Rozanska M, Gosselin M, Poulin M, Wiktor JM, Michel C (2009) Influence of environmental factors on the development of bottom ice protist communities during the winter and spring transition. Mar Ecol Prog Ser 386:43–59

Schünemann H (2004) Studies on the Arctic pack-ice habitat and sympagic meiofauna: seasonal and regional variabilities. PhD Thesis. Christian-Albrechts Universität Kiel

Spindler M (1996) On the salinity tolerance of the planktonic foraminifera, pp 85–91

Thomas DN, Dieckmann GS (2000) Sea Ice, Wiley, Blackwell

Weeks W (2010) *On sea ice*, University of Alaska Press, Alaska

Wiktor J (2015) *Morskie pierwotniaki Arktyki* Rozprawy i Monografie 24. Institut of Oceanology Polish Academy of Science, Sopot, pp 177. *in polish*

The Impact of Small-Scale Fisheries Activities Toward Fisheries Sustainability in Indonesia

Nisa Ayunda, Mariusz R. Sapota and Anna Pawelec

Abstract Indonesia is an archipelago country, which has big fisheries resources. The fisheries sector has contributed 3% of *gross domestic product* (GDP) in 2012. More than 80% of Indonesian fish catches are from small-scale fishery. Hence, it is crucial to investigate the performance of small-scale fisheries in relation to the sustainable fish exploitation. The purpose of this study is to calculate fisheries degradation rate to show the resource exploitation trend in Indonesian small-scale fisheries. We focus on small-scale fisheries that use hook and line and gillnet to catch Eastern little tuna, Skipjack tuna, Red snappers, Blue line sea bass, and Halibut. The fisheries degradation rate shows the fish resources depletion caused by fishing. The fisheries degradation rate had trend to reach *the threshold score* with the average coefficient of 0.33 in ten years. This means that Indonesian small-scale fisheries lead to high fish exploitation. We recommend that the Indonesian government improve regulation of the small-scale fishing activities by limiting the fishing days and gears used, so the activities become sustainable.

Keywords Degradation rate · Indonesia · Small-scale fisheries
Sustainable fishing efforts · Sustainable fisheries

1 Introduction

More than 80% of Indonesian fish catches are from small-scale fisheries (DJPT 2013, 2015). Generally, fishermen apply simple technics and are fishing in coastal area. Their activities are dominated by non-motor boats and outboard motor boats (Adhuri 2013). Fishermen usually fish in daily basis (Mochtar 2016)—start in the early morning and finish by late afternoon or start late in the night and finish in the early morning. They usually work in pair in one boat. Their basic knowledge on

N. Ayunda (✉) · M. R. Sapota · A. Pawelec
Faculty of Oceanography and Geography, Institute of Oceanography,
University of Gdansk, Al. M. Pilsudkiego 46, 81-378 Gdynia, Poland
e-mail: nisa.ayunda@phdstud.ug.edu.pl; nisa.ayunda@phstud.ug.edu.pl

T. Zielinski et al. (eds.), *Interdisciplinary Approaches for Sustainable Development Goals*, GeoPlanet: Earth and Planetary Sciences,
https://doi.org/10.1007/978-3-319-71788-3_11

fishing technics is based on their ancestor experiences (Ayunda 2014; Adhuri 2013). Because of family characteristics, Indonesian small-scale fisheries have a limited regulation (Indonesian Fisheries Law No. 45/ 2009).

Indonesia (Fig. 1) has big fisheries resources. Before 1998, in Indonesia, there were only a few regulations in fish resources exploitation. During the reformation stage in politics, Indonesia government has focused more strongly on developing this sector to support national income (KKP 2014). Many policies have been applied to reform fishery rules until 2014, when the newly elected president has published a new doctrine "Indonesia as the World Maritime Axis" to stimulate the acceleration in the marine development (Kominfo 2016). To achieve it, the government has used 5 pillars, including: revitalization the maritime culture, managing maritime resources for food security, developing maritime infrastructure, improving maritime diplomacy, and boosting maritime defense capacity (Kominfo 2016). To adopt this new dogma, Marine Affairs and Fisheries Ministry has developed three concepts in sovereignty, sustainability, and prosperity (KKP 2014). In line to the new principle, Indonesia government has a priority to eradicate *illegal, unreported and unregulated* (IUU) fishing activities (KKP 2014).

To manage the fish exploitation, Indonesia implemented *Total Allowable Catch* (TAC) based on the National Fisheries Law No. 45/2009. The government evaluated TAC estimation up to 90% of *maximum sustainable yield* (MSY) (Triyono 2013; Utami 2015). However, there is still debate on using the MSY and TAC in Indonesia (Triyono 2013). Hence, it is crucial to manage the small-scale activities towards more sustainable fisheries resource use. The rules that regulate the sustainable use of fisheries resources are still under development in Indonesia. Indonesia government applies MSY as a tool to manage fish exploitation and estimate the fisheries potential based on Gordon-Schaefer model (Pet et al. 2005; Wiadnya and Halim 2008). Pet et al. (2005) and Wiadnya and Halim (2008) evaluated that MSY was unsuitable as management target, especially for Indonesia.

Fig. 1 The Map of Indonesia (*Source* worldtravel.com)

They found many weaknesses in this analysis including the quality of Indonesian statistics data. The method calculated fish stock in equilibrium and catch per effort as a good indicator for the size of the fish stock, and many Indonesian fishery scientists and policy makers misinterpreted the output.

Since 2000, the Indonesian government has attempted to regulate small-scale fisheries by assisting and empowering *local institution* (e.g., *Panglima Laot*, *Awik-awik*, *Sasi*, etc.) (Ayunda 2014; Adhuri 2013; Ardhianto 2011; Hidayat 2005; Adhuri and Indrawasih 2003) and prevent *illegal* fishing boats from neighboring countries (KKP 2014). The program was unsuccessful because of the lack of cooperation and low commitments of the actors involved in management. For instance, the establishment and implementation of *Awik-awik* in East Lombok (Ayunda 2014; Ayunda et al. 2014) failed to manage the fisheries resources for longer than time of government support (5 years). Even though that one of the goals is to prevent the IUU fishing, there were fishermen that still involved in such activity (Ayunda and Anna 2015; Ayunda 2014).

Food and Agricultural Organization (FAO) recorded that about 90% of fisheries production in the World came from the small-scale fisheries activities (FAO 2015). FAO (2009) defined the small-scale fisheries as "*fishing activities that used minimal amount of capital, low level technologies and household-unit entities*".

In Indonesia, small-scale fisheries have an unclear definition. Referring to the new Indonesia regulation (No. 7/2016 about *Protection and Empowerment of Fishermen, Fish Raisers and Salt Farmers*), Indonesian small-scale fisheries are fishermen who catch fish for daily needs, without or with fishing vessels less than 10 gross tonnage (GT). While *Indonesian Fisheries Law* No. 45/ 2009 gave small-scale fisheries concept as fishermen who are fishing for their daily needs without or with boats under 5 GT. Unclear definition of small-scale fisheries, and undefined amount of daily needs catches make it difficult to track it in the national data (DJPT 2013, 2015). Even FAO (2014) had published that Indonesian small-scale production reached to 95% of the total national fisheries.

In the Baltic Sea, after stipulation of the MSY as a tool, the improvement in fishery management was noticed. One of the successes is recovering the Eastern Baltic Cod, which until the mid-2000s was overfished (Eero et al. 2015; OCEANA 2012). The harvesting control by setting TAC also contributed to the recovery of its fisheries (Eero et al. 2012).

In this study, we investigate the performance of Indonesian small-scale fisheries in relation to the sustainable development. We will measure the performance by calculating fisheries degradation rate in Indonesia by using MSY as a benchmark. We distinguish Indonesian small-scale fisheries which catch Eastern little tuna (*Euthynnusaffinis*), Skipjack tuna (*Katsuwonuspelamis*), Red snappers (*Lutjanus-saguineus*), Blue lined seabass (*Epinephelus*), and Halibut (*Psettodeserumei*) using hook and line and gillnet.

We describe the trend of Indonesian small-scale fleets and how its influences into the fisheries resources through annual fisheries statistic data from 2004 to 2014.

2 The Framework: Resource Sustainability in Small-Scale Fisheries

The goal of fisheries management is maintaining the fish exploitation into sustainable to accomplish benefit in biology and economics not only in short term but also in long term (Cochrane 2002; FAO 1995). In the *organization for economic cooperation and development* (OECD), the examination of the biology and social-economic situation are the important indicators in the assessment of sustainable management in fisheries sectors (Gallic 2002).

Using MSY, many researchers, notably in Indonesia, have developed degradation rate to evaluate the changing of fisheries resources caused by fishing activities for human consumption, pollution from urban area, etc. (Najamuddin et al 2016; Kurniawan 2015; Fauzi and Anna 2005; Adhuri and Indrawasih 2003; Anna 2003; Turner et al. 1999).

Concerning our understanding, we assume that fishery sustainability management could be investigated by looking at the biological, economics, and technological state of the fishing activities, by using sustainable yield, sustainable economic yield and efficiency of technology as the benchmark (Fig. 2). We find that *biological parameters* (e.g., actual catch, and fisheries degradation rate), and *economics* (e.g., actual profit, and fisheries depreciation rate) are below sustainable level, while the *technology usage* (e.g., efficiency of actual technology used) is in the efficient level (*symbol* +). Together they indicate that implementation of fisheries management is capable of managing fish exploitation into the sustainability. On the other hand, if *biological* and *economic states* are approaching to or are higher than sustainable level, and *technology* is inefficient (symbol −), they indicate

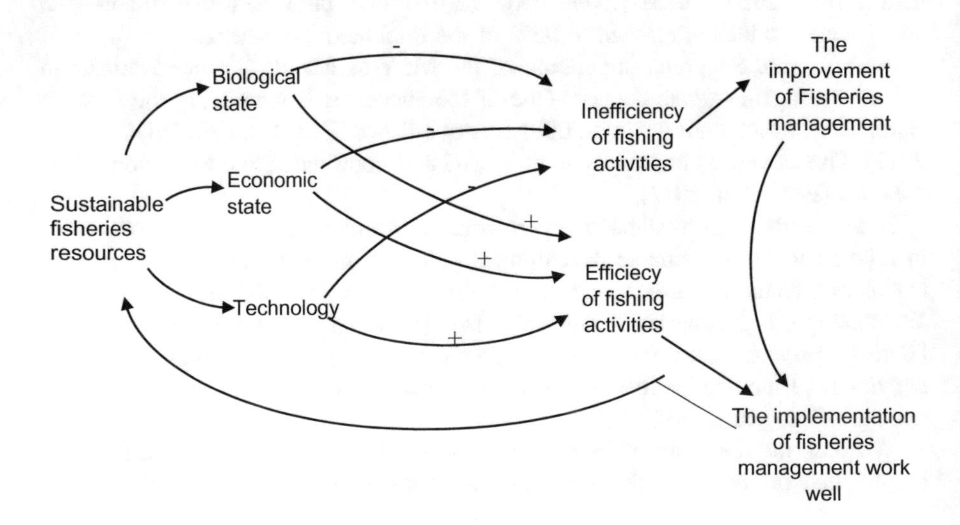

Fig. 2 Our research flowchart

that fisheries management is incapable of governing fisheries utilization into the sustainability, so the enforcement of fisheries management need to be improved.

Following our research (Fig. 2), we focus on *biological state* analysis by calculating degradation rate, described by the ratio between actual and sustainable catch (calculated by the fishing activities). The measurement was adapted by Amman and Duraiappah (2002) in their research of the land resources degradation in Kenya. They gave an example calculation of land degradation between two agents in the land used.

In the fisheries, this concept could be comprehended by the interaction between sustainable catch and actual catch (Anna 2003) (Eq. 1). In this calculation, some of fishery scientists agreed that 0.5 is *the threshold score* (Najamuddin et al. 2016; Ayunda et al. 2014; Fauzi and Anna 2005; Anna 2003). It indicates that fishing activities have caused 50% decrease of fisheries resources. So, if the annual degradation rate rises to 0.5, it shows that fishing activities have caused high exploitation and the management requires developing to be capable to manage the sustainability.

3 Methods

For the purpose of this study, we use secondary data on Indonesian annual fisheries data published by Ministry of Marine Affair and Fisheries of Indonesia through Indonesia Directorate General of Capture Fisheries. We use the data on *Capture Fisheries Statistics of Indonesia 2010* and *Statistics of Marine Capture Fisheries by Fisheries Management Area* (FMA) to estimate the small-scale fisheries activities from 2004 to 2014.

In the reports, the fisheries data is derived from the region data. This measurement assumed that the vessels have been operated in their coastal water territory. For the larger vessels, the production would be added in their landing port. This statistics has the substance information of the fisheries production (landing weight), the number of boats and fishing gears in fisheries activities in Indonesia waters (both in marine and inland area), and aquaculture activities.

In this study, we focus on substantial catch in pelagic, demersal, and reef (coral) fisheries based on the annual report (e.g., Eastern little tuna, Skipjack tuna, Red snappers, Blue lined seabass, and Halibut). From fishing gears, we select hook and line and gillnet used on non-motor and outboard machine boats. We focus on the fleets because many sustainable fisheries were addressed to these activities (FLO 2014; Ayunda 2014; Ayunda et al. 2014; Banks and Lewis 2011). The small-scale fisheries data both in production and gears are selection from the annual fisheries data following function that presented in *data analysis*.

The secondary data were collected using some publications that are related to fisheries management in Indonesia (Najamuddin et al. 2016; Kurniawan 2015; FLO 2014; Adhuri 2013; Ardhianto 2011; Bank and Lewis 2011; Fauzi and Anna 2005;

Hidayat 2005; Adhuri and Inderawati 2003; Anna 2003), and also our previous research (Ayunda 2014; Ayunda et al. 2014; Ayunda and Anna 2015).

3.1 *Data Analysis*

Fisheries degradation rate is part of our investigation on biological state in our research flow (Fig. 2). For the purpose of this study, we adapted the measurement by Amman and Duraiappah (2002) (Nadjamudin et al. 2016; Kurniawan 2015; Ayunda 2014; Fauzi and Anna 2005; Anna 2003). The calculations follow the equation:

$$\varnothing_D = \frac{1}{1 + e^{\frac{h_s}{h_a}}} \tag{1}$$

where:

$\varnothing_D$ coefficient of fisheries degradation rate
h_s sustainable catch (ton)
h_a actual catch (ton)

The equation shows that the coefficient of degradation rate is exponential between sustainable catch and actual catch. The result of maximum degradation coefficients is 1. It is usually approached when one of the variables shows in a negative value. It indicates that fish exploitation has caused 100% decrease of quality and quantity of fisheries resources and this situation is undesirable in the fisheries management. So, *the threshold score* of this calculation is mentioned to be 0.5 (Fauzi and Anna 2005; Anna 2003). This score indicates that fishing activities have caused 50% decrease of fisheries resources. It is a result when the exponential of actual and sustainable catch is 1 or when sustainable catch shows 0. Thus, by existing of the management, fisheries degradation rate should be below *the threshold score*. If the fisheries degradation rates increased to *the threshold*, the fish exploitation triggered high declining of quality and quantity of fisheries resources (Kurniawan 2015).

3.1.1 Biology Variables

Applying the equation above, we need to calculate biology variables (e.g., actual catch (h_a) and sustainable catch (h_s)). Actual catch (h_a) is the total of decompositions of fish production of five species (h_{ijt}) using hook and line (l) and gillnet (k) (Anna 2003) following the equation below:

$$h_a = \sum_l \sum_k h_{ijt} \quad i = 1, 2, 3, 4, 5 \tag{2}$$

Decomposing of five species fish production (h_{ijt}) is calculated by the following equation:

$$\begin{aligned} h_{ijt} &= \varnothing_i h_{it} \\ \varnothing_i &= \prod_{t=1}^{m} \left[\frac{h_{ij}}{\sum h_i} \right]^{\frac{1}{n-1}} \\ h_{ijt} &= \left[\prod_{t=1}^{m} \left[\frac{h_{ij}}{\sum h_i} \right]^{\frac{1}{n-1}} \right] * h_{it} \end{aligned} \tag{3}$$

The formula involves proportion of fisheries production species i by gear j ($\varnothing_i$), fisheries production species i in time t (h_{it}) and sum of fish production caught by gears (h_i) in time t. h_i is decomposed from national fisheries production following the formula:

$$h_i = \left[\prod_{t=1}^{m} \left[\frac{h_{it}}{h_{lt} + h_{kt}} \right]^{\frac{1}{n-1}} \right] * h_t \tag{4}$$

This equation engages total production of species i (h_{it}), total production of hook and line (h_{lt}), and total production of gillnet (h_{kt}) in time t. Due to the unavailability of the exact data on total production by hook and line and gillnet, we estimate these data using the following equation:

$$\begin{aligned} h_{lt} &= \left[\prod_{t=1}^{m} \left[\frac{l_t}{g_t} \right] \right] * h_t \\ h_{kt} &= \left[\prod_{t=1}^{m} \left[\frac{k_t}{g_t} \right] \right] * h_t \end{aligned} \tag{5}$$

where:

- i Eastern little tuna, Skipjack tuna, Red snappers, Blue lined seabass, and Halibut
- j fishing gears (longline and gillnet)
- n the number of gears that used in calculation
- h_t national fish production in time i (ton)
- h_a actual catch in time t (ton)
- h_{it} production of species i in time t (ton)
- h_{ij} production species i caught by gear j (longline or gillnet) (ton)
- h_i disaggregation of fish production of species i by gears (longline and gillnet) (ton)

$Ø_i$ proportion of species i caught by gear j
h_{ijt} decomposition of species i in time t caught by gear j (ton)
h_{lt} fish production caught by longline in time t (ton)
h_{kt} fish production caught by gillnet in time t (ton)
g_t total gears in time t (unit)
f_t total longline in time t (unit)
k_t total gillnet in time t (unit)

Anna (2003) suggested applying geometric mean (Π) in time series data to calculate proportion of data accurately.

Sustainable catch (h_s) as potential fisheries production is calculated by following the surplus production of CYP model (Clarke et al. 1992). In this function, we assume that biomass growth (x) without any interference from human activities (Fauzi and Anna 2005) follows the equation:

$$\frac{\partial x}{\partial t} = rx\left(1 - \frac{x}{K}\right) \tag{6}$$

We assume that the potential fisheries production (h_s) is dependent on the fishing effort (E_t), the biomass growth (x), and the catchability coefficient (q) following the Cobb-Douglas equation (Najamuddin et al. 2016; Kurniawan 2015; Fauzi and Anna 2005):

$$h_s = qxE_t \tag{7}$$

When we calculate (Eq. 7) in equilibrium $\left(\frac{\partial x}{\partial t} = 0\right)$, we could substitute ($x$) in this equation into (Eq. 6), and the potential fisheries production becomes:

$$h_s = qKE_t\left(1 - \frac{q}{r}E_t\right) \tag{8}$$

In this formula, potential fisheries production contains the biology parameters, e.g., intrinsic growth rate of fisheries (r), catchability coefficient (q), and carrying capacity (K). We calculate r, q, and K by the following equation (Najamuddin et al. 2016; Ayunda 2014; Fauzi and Anna 2005; Clarke et al. 1992):

$$ln(U_{t+1}) = \frac{2r}{(2+r)}ln(qK) + \frac{(2-r)}{(2+r)}ln(U_t) - \frac{q}{(2+r)}(E_t + E_{t+1}) \tag{9}$$

where:

U_t catch per unit effort (CPUE) in time (t)
U_{t+1} catch per unit effort (CPUE) in time +1
E_t fishing effort in time t (fishing days)
E_{t+1} fishing effort in time + 1 (fishing days)
h_a actual catch in time t (ton)
r intrinsic growth rate of fisheries

q catchability coefficient
K carrying capacity (ton)

where:

$$U_t = \frac{h_a}{E_t} \tag{10}$$

If $\frac{2r}{(2+r)} = a, \frac{(2-r)}{(2+r)} = b, \frac{q}{(2+r)} = c$, so coefficient r, q, and K could be calculated by following the equation:

$$\begin{aligned} r &= \frac{2(1-b)}{(1+b)} \\ q &= -c(2+r) \\ K &= \frac{e^{\frac{a(2+r)}{2r}}}{q} \end{aligned} \tag{11}$$

3.1.2 Standardization Fishing Effort

We estimate the number of hook and line and gillnet used in Indonesian small-scale fisheries. We calculate the fishing effort by multiplying the numbers of days in the sea (fishing days) and the numbers of estimated gears. In the fishing effort measurement, we calculate gear standardization because hook and line and gillnet have differences in ability to catch fish, by the following the equation (Najamuddin et al. 2016; Kurniawan Kurniawan 2015; Fauzi and Anna 2005):

$$\begin{aligned} E_t &= \varphi_{it} D_{it} \\ \varphi_{it} &= \frac{U_{it}}{U_{jt}} \end{aligned} \tag{12}$$

where:

E_t effort standardized
φ_{it} index of fishing effort gear i at time t
D_{it} the number of days of fishing gear i at time t
U_{it} catch per effort (CPUE) of fishing gear i at time t
U_{jt} catch per effort (CPUE) of fishing gear is used as the standard

We choose the standard fishing gear which has high CPUE by high CPUE indicates that the fishing gear catches the fish more efficiently (Kurniawan 2015).

We apply Scatter plot graph and Pearson correlation to validate our selected data. The result of Pearson correlation is the coefficient of correlation (r) with ranges from −1 to +1. The high relation between two variables is shown by the value of r close to +1 or −1 (CRS 2015).

4 Results

Indonesian fisheries, including small-scale ones, usually capture multispecies. The valuable fish catches in tuna fisheries, coral reef fish, and demersal fish are: Eastern little tuna (had landed value 1173.16 USD per ton), Skipjack tuna (1193.00 USD per ton), Red snappers (2258.11 USD per ton), Blue lined seabass (444.05 USD per ton), and Halibut (213.03 USD per ton) [1 USD = 11835 Indonesian Rupiah (IDR) (OANDA 2017)] in 2014 (DJPT 2015). Apart from high landed values, the species are also found nearby Indonesian coastal waters. The small-scale fisheries in Indonesia are dominated by fishermen who use up to 12 m in overall length boats with below 4 horse power (HP) in engine, and under 5 GT in capacity. They usually catch fish nearby coastal area.

4.1 Actual Catch

The results of actual catch (h_a) of Indonesian small-scale fisheries reached the highest amount of 80 817.50 ton in 2014 and the lowest was 4,023,403 ton in 2005 (Fig. 3). The rise of tuna fisheries during last decades contributed increasing of actual catch. If the exploitation remains unregulated, it could cause a decline of tuna fisheries production similar to other species. High level of fishing effort of red snappers, blue line seabass and halibut is one of main reasons of the decline of production in the long term (Najamudin et al. 2016; Ayunda 2014; Anna 2003).

The calcualtion showed that actual catch had a correlation with fisheries production following regression $y = 0.0212x - 46485$ with correlation coefficient 0.96 [higher than r table (0.521) significant 0.05] and p-value 0.000 (Fig. 4). It provided

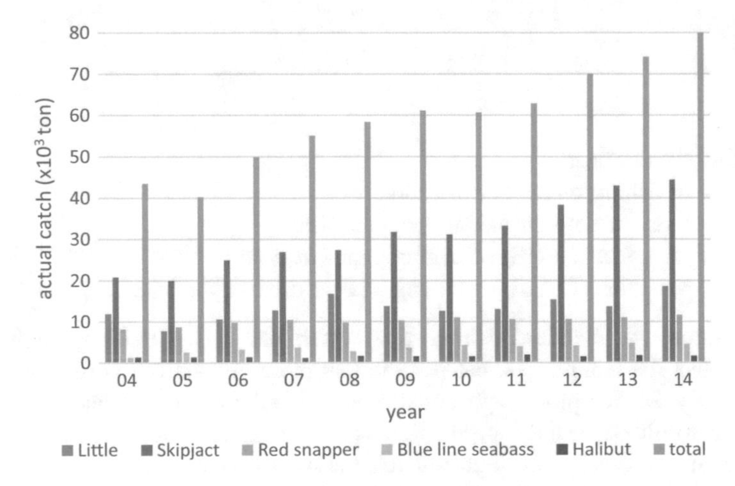

Fig. 3 Estimation of Indonesian small-scale actual catch 2004–2014

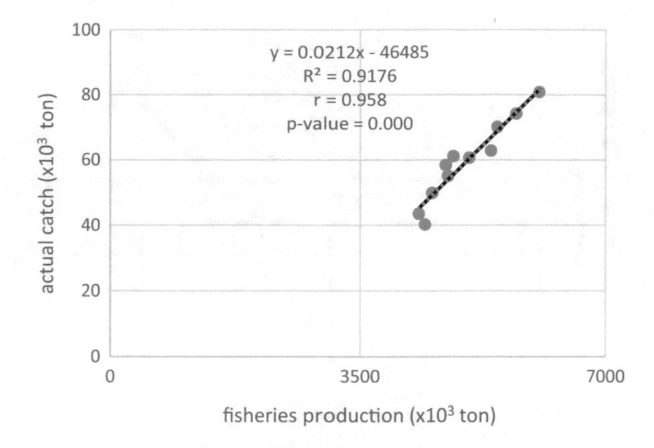

Fig. 4 The validation of estimation of Indonesian small-scale actual catch and national fisheries production

that the selected data of actual catch representing a sample of small-scale production statistically.

4.2 *Standardization Fishing Effort*

In standardization fishing effort, we estimate fishing effort of small-scale fisheries by multiplying the numbers of days in the sea and the numbers of exposed gears. The estimated numbers of gears showed that the highest use of hook and line and gillnet was in 2007 (Fig. 5). The data showed the correlation between the estimated gear number used and total hook and line and gillnet that operated in Indonesia was

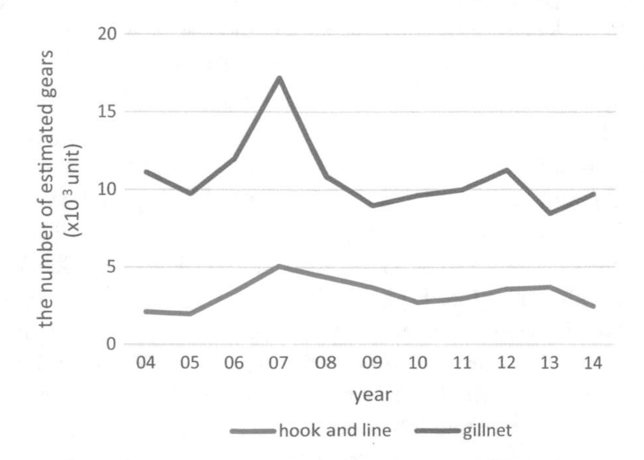

Fig. 5 The estimated number of gears of Indonesian small-scale fisheries in 2004–2014

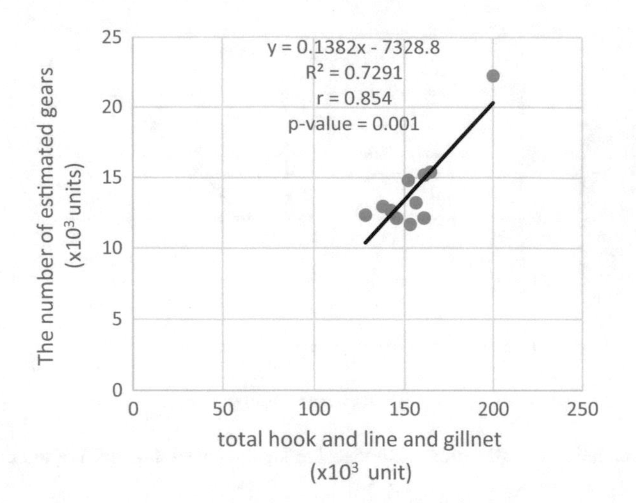

Fig. 6 The validation of the number of estimated gears and hook and line and gillnet

following the regression $y = 0.14x - 7329$ by coefficient 0.85 [higher than r table (0.521) significant 0.05] and p-value of 0.001 (Fig. 6). It confirmed that our estimated gears numbers describe a sample of hook and line and gillnet operated by small-scale fisheries.

Referring to our previous research (Ayunda 2014; Ayunda et al. 2014), we calculated the number of days in sea at 200 days per year. Small-scale fishermen (e.g., East Lombok) have a custom of one day free (on Friday) during a week and a few of them take a break on January and February because of high tides.

The fishing effort (E_t) reached the maximum of 4,190,827 fishing days in 2007 and the minimum of 2,233,005 fishing days in 2005. High number of operating hook and line and gillnet triggered the highest fishing effort in 2007. Catch per unit effort (CPUE) as the measurement of density of exploitation stock presented a tendency to rise every year (Table 1). The fishing effort calculation had a correlation with the selected gears followed regression $y = 151.06x + 1E^{+06}$ with Pearson correlation (r) in 0.75 and p-value in 0.009 (Fig. 7).

Applying Eq. 9, we need to validate Ln $CPUE_{t+1}$; Ln $CPUE_t$, and $E_t + E_{t+1}$. The result is that the correlation between Ln $CPUE_{t+1}$ and Ln $CPUE_t$ (Fig. 8) follows the regression $y = 0.83x - 0.62$ with coefficient 0.735 [higher than r table (0.521) significant 0.05] and p-value 0.02. While Ln $CPUE_{t+1}$ and $E_t + E_{t+1}$ (Fig. 9) showed correlation following the regression $y = -0.14x - 3.07$ by coefficient −0.522 [in point of r table (0.521) significant 0.05)] and p-value of 0.122. The negative value represented that high Ln $CPUE_{t+1}$ triggered lower $E_t + E_{t+1}$.

Regarding the bio-economics model that applied in this analysis, the actual catch (h_a) and fishing effort (E) are the total of five species observed and sum of fishing effort that have been standardized. As a result, the intrinsic growth of fisheries population (r) was 0.26; catchability coefficient (q) was 0.0000000530; and carrying capacity (K) was 828,185.29 ton (Table 2). We applied these parameters to

Table 1 The standardization of fishing effort (E_t) of Indonesian small-scale fisheries

Year	Selected gears (unit)		Index gillnet (φ_{it})	Fishing effort (E_t) (day fished)	Catch per effort unit (CPUE)
	Line	Gillnet			
2004	2095	11114	0.63620	2823162	0.01541
2005	1963	9723	0.68143	2233005	0.01802
2006	3404	11979	0.95902	3574153	0.01396
2007	5068	17176	0.99589	4190827	0.01314
2008	4346	10840	1.35303	3935487	0.01483
2009	3680	8986	1.38208	3622274	0.01688
2010	2719	9626	0.95338	3212018	0.01888
2011	2972	9987	1.00433	3250503	0.01933
2012	3566	11240	1.07084	3604185	0.01944
2013	3692	8450	1.47457	2422835	0.03062
2014	2448	9674	0.85422	2945759	0.02744

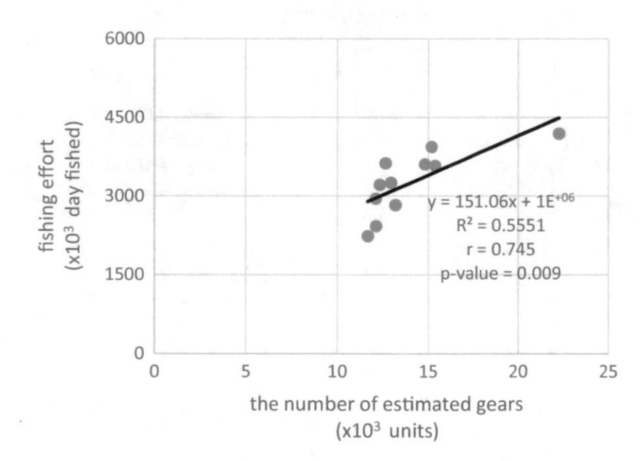

Fig. 7 The validation of fishing efforts and the number of estimated gears

estimate sustainable catch. The highest of sustainable catch was 52,332.29 ton in 2014, and the lowest was 28,499.89 ton in 2007 (Table 3).

Following the analysis, we presented the graph of sustainable catch and fishing effort. It showed that sustainable catch had a correlation with fishing effort following the equation $y = -9E^{-09}x^2 + 0.044x\text{-}3E^{-13}$. The negative value means that rising fishing effort tiggered a decreasing sustainable catch (Fig. 10).

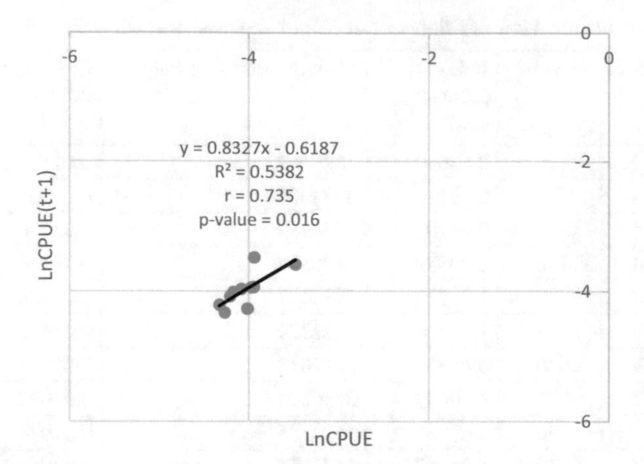

Fig. 8 The validation of Ln $CPUE_{(t+1)}$ and Ln $CPUE_{(t)}$

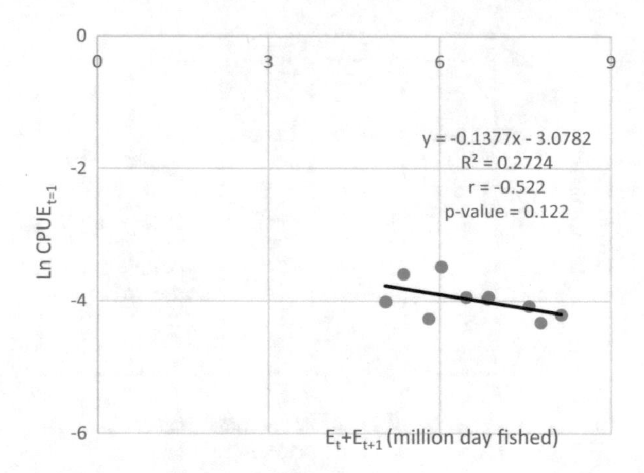

Fig. 9 The validation of E_t+E_{t+1} and Ln $CPUE_{t+1}$

4.3 The Fisheries Degradation Rate

The calculations described above have shown that the highest coefficient of fisheries degradation rate was 0.38 in 2007, the lowest was 0.21 in 2005, and the average score was 0.33 (Fig. 11). The resulting annual coefficient was below the threshold with a tendency to rise according to the following equation: $y = 0.0096x + 0.25$. The increasing score exhibited high exploitation from Indonesian small-scale fisheries. The high rate also indicates that actual catches are rising and are higher than the sustainable level every year (Fig. 12).

Table 2 Biology parameters calculations of Indonesian small-scale fisheries 2004–2014

Biology parameters	Value
r (intrinsic growth rate of fisheries)	0.26
q (catchability coefficient)	0.0000000530
K (carrying capacity (ton))	828185.29
R^2	0.74

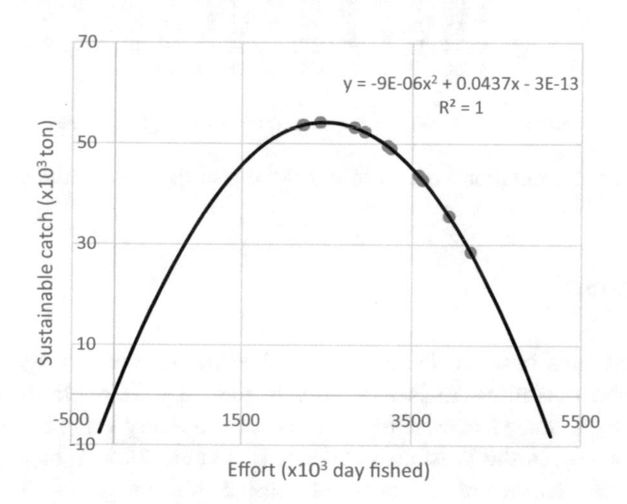

Fig. 10 Fishing effort (E_t) and sustainable catch (h_a)

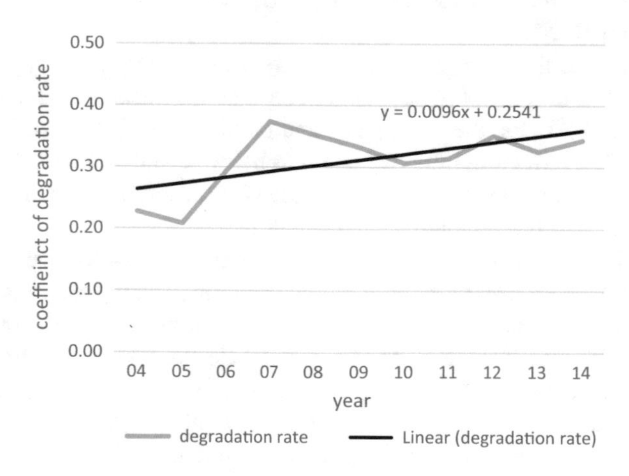

Fig. 11 Fisheries degradation rate of Indonesian small-scale fisheries 2004–2014

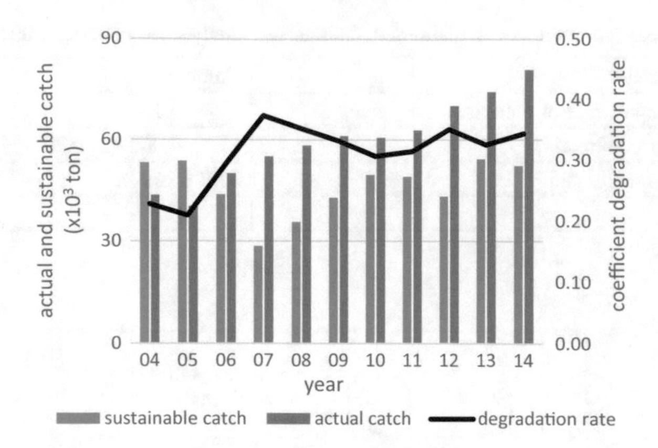

Fig. 12 Actual catch, sustainable catch and fisheries degradation rate of Indonesian small-scale fisheries 2004–2014

5 Discussion

This study presents how the Indonesian small-scale fisheries changes in terms of production and operations in last decade. In the early 2000, Indonesian fishery scientists have estimated overfishing because of too many gears in use in several fisheries resources, in the Western Indonesia (Pet et al. 2005; Wiadnya and Halim 2008). By this failure in catches, the sustainable programs for Indonesian small-scale fisheries are centered in coral reef triangle area (East part of Indonesia) in the last decade (FLO 2014; Ayunda 2014; Ayunda et al. 2014; Adhuri 2013; Adrianto 2011; Banks and Lewis 2011), and it covers 45% of national fisheries production (DJPT 2015).

To estimate the fisheries potential, Indonesia government has applied MSY as the standard. Pet et al. (2005) recommended that we needed to be more careful, when applying MSY into the multi gears and multi species fishing activities such as in Indonesia. The reduction of actual catch to below sustainable catch, in reality, is very difficult. Actual catch could be influenced by the amount of fish, the selecting fishing ground, and the regulation (Gallic 2002).

The calculation of fish exploitation is one of crucial importance so the fishing activities could give a benefit to biology and economy. Many scientists have argued that MSY as the standard is the conventional calculation (Pet et al. 2005; Wiadnya and Halim 2008), but it is much easier to calculate than other benchmarks (Gallic 2002). Many calculations and managements are developed from this basic measurement, for example, the fisheries degradation rate. Fisheries degradation rate could be a tool to evaluate how fishing activities influence the resources in biology and economy by calculating changes of actual catch to the sustainable level.

The results of degradation rates calculation between 2004 and 2014 in Indonesian small-scale fisheries showed that the average of annual scores of coefficient degradation was 0.33. Average score is still below *0.5*, but when we analyze the results further, the annual scores have a tendency to reach *the threshold score*. In this state, the small-scale catches have caused high percentage in exploitation of fisheries resources. So, Indonesia government need to improve the enforcement of management to be more sustainable.

Indonesian government has implemented several polices to manage small-scale exploitation for sustainability. Extensively empowerment of *local institution* to govern coastal fisheries resources, e.g., *Awik-awik* in East Lombok and West Lombok Regions, *Sasi* in Mollucas, *Panglima Laut* in Aceh, and many other regions in early 2000s (Ayunda 2014; Ayunda et al. 2014; Adhuri 2013; Adrianto 2011; Adhuri and Inderawati 2003). But, until now, many interferences still appear to disincentive these performances, e.g., lack of cooperation between stakeholders and low incentive between actors who were involved in *Awik-awik institutional*, East Lombok Region (Ayunda 2014; Ayunda and Anna 2015).

Awik-awik in East Lombok has been developed since many irresponsible fishing activities appeared near East Lombok costal region during the 1990s. It was initiated by a local small-scale group, called *Nautilus* (Ayunda 2014; Syaifullah 2009). At the beginning, it was only an unwritten agreement between local fishermen, but since 2003, it has been a guidance book of management coastal area under *Awik-awik* supported by government (national, regional, and local), education institution, research institution, and local *non-government organization* (NGOs). The guidance book has regulated the quality of fish production (e.g., banning catch baby fish and mature fish); fishing effort (e.g., zonation of fishing activities, restricting *fishing additional devices* (FAD) such limitation of light used); and surveillance and monitoring (e.g., giving sanctions for each violation). It still failed to regulate selecting gears (e.g., the number of hook and its size which is allowed to operate); landing weight quota; limiting the fishing days, so the activities seemed to have a tendency to be *Open Access* (Table 4).

Table 3 Sustainable catch of Indonesian small-scale fisheries from 2004 to 2014

Year	Fishing effort (E) (day fished)	Actual catch (h_a) (ton)	Sustainable catch (h_s) (ton)
2004	2823162	43510.12	53202.36
2005	2233005	40234.03	53686.35
2006	3574153	49897.15	43716.59
2007	4190827	55063.49	28499.89
2008	3935487	58371.84	35613.05
2009	3622274	61146.21	42770.13
2010	3212018	60650.03	49530.84
2011	3250503	62824.47	49022.64
2012	3604185	70062.92	43130.68
2013	2422835	74191.05	54199.92
2014	2945759	80817.50	52332.29

Table 4 Advantage and disadvantage of *Awik-awik* in East Lombok

The regulation	Strengthens	Weakness
Quality of catch	Selected catch	Regulate selecting gear
		Quota of production
Effort	Selected FAD	Limiting the fishing days
	Zonation	Fishing license
Monitoring and surveillance	Sanction for each violation	Low incentive

Following our study, the improving of the regulation could be implemented to other region. Such other Indonesia coastal areas (e.g., in Maluku and Sulawasi islands), government has cooperated with some of NGOs in Indonesian such *Action Plan Indonesian Tuna Fisheries* from 2009 (Banks and Lewis 2011) and *Indonesia Tuna Fairtrade certification* from 2013 (FLO 2014) to manage small-scale tuna fisheries. The program offer managing selected gears, improving the landing quota and quality, raising fish price by about 10%, banning catch and assassinate endangered species (e.g., mahi-mahi, dolphin, sea turtles, etc.), and also introducing the fishing license.

Reducing the number of fishing days could be difficult to be implemented, because most of Indonesian small-scale fishermen provide food directly for their household. Indonesian government has cooperated with education institutions, research institutions, and local society to apply reducing the fishing days by introducing and assisting the *Marine Protected Area Program* (MPA) based on *local institutional* in 2000s (e.g., the North part of East Lombok, West Lombok, Thousand Islands, Bunaken National Park, etc.), but it also still showed inefficiency because the low incentive to follow the regulation (Ayunda 2014; Ardianto 2011; Christie 2004; Clifton 2003). Christie (2004) published his finding at *American Fisheries Society Symposium* 42 that 4 location of MPA in Southeast Asia (Salvador Island, Twin Rocks and Balicasag (the Philipines); Bunaken National Park (Indonesia)) succeed to manage the biology but still failed in social. This disincentive could be managed by participating small-scale fishermen in ecotourism and marine aquaculture (*mariculture*), some parts of the Indonesian coastal areas are developed to tourism and *mariculture* because of the clear waters and *Coral Reef* Sea based.

In this study we show that the positive influences of Indonesian small-scale fisheries management is present in specific regions and species of fish but not yet in national level. Therefore, it needs to be developed in more regions, more species, particularly coral and demersal fish (e.g., red snappers and blue lined seabass, etc.), and more engagement between the actors (e.g., *local institutions*, government, research institution, education institution and NGOs), so the influences could be also visible in national data of annual fisheries.

In this study, we focused on fisheries data from 2004 to 2014 because annual data following 2014 are unavailable overtly yet; the program (notably *Tuna fisheries management*) is still in progress in a few regions in Indonesia (FLO 2014). Thus, its impact could not be evaluated in this study. Therefore, assimilating of degradation rates could also encourage to improve information of fisheries exploitation level for developing annual fisheries data and the subsequent research.

6 Conclusion

We investigated the performance of Indonesian small-scale fisheries in relation to the sustainable catches by calculating fisheries degradation rate. Fisheries degradation rate presents depletion levels of fisheries resource in Indonesian small-scale activities. The annual coefficients rise from 2004 to 2014, and actual catches are approaching or are higher than the sustainable catch. It determined high fisheries exploitation. In last two decades, many programs have been employed to manage Indonesian small-scale fisheries, in a few part of Indonesian coastal area. Unfortunately, many interferences still appear to disincentive these performances.

This study suggested to improve the enforcement of management by increasing the cooperation between Indonesia government and education institutions, research institutions, *non-government organizations* (NGOs), also local fishermen by driving limiting the fishing effort, and improving the process of gathering of annual fisheries data. The fishing effort limitation could be determined by introduction of licenses for small-scale fishing in more regions and nationwide or by reducing the number of fishing days in favor of encouraging tourism and *mariculture* as the source of income in rural areas, so the management has been impacted also to national data. Fisheries degradation rate, calculated as presented here, should be introduced as standard, common means of measurement of trends in regional and national institutions responsible for fisheries resources.

References

Adhuri DS (2013) Opportunities and challenges for ecosystem based fisheries management in East Lombok: some impressions from a short field visit

Adhuri DS, Indrawasih R (2003) Pengelolaan Sumber Daya Alam Secara Terpadu (Co-Management Sumber Daya Alam) Pembelajaran dari Praktek Pengelolaan Sumber Daya Laut di Bangka-Belitung, Jawa Tengah, dan Jawa Timur serta Pengelolaan Taman Nasional Lore di Sulawesi Tengah (Co-management in natural resources: lesson learn from applied sea resources management in Bangka-Belitung, Central Java, East Java, and Central Sulawesi). Jakarta: Puslitbang Kemasyarakatan dan Kebudayaan-LIPI

Adrianto L (2011) Konstruksi Lokal Pengelolaan Sumberdaya Perikanan Di Indonesia (Local construction of fisheries resources management). IPB Press, Bogor

Amman HM, Duraiappah AK (2002) Modeling instrumental rationality, land tenure and conflict resolution. Comput Econom 18:251–257. Kluwer Academic Publishers: The Netherlands

Anna S (2003) Model *Embedded* Dinamik Ekonomi Interaksi Perikanan-Pencemaran (Embedded model in economics dynamic in interaction between fisheries and pollution). Desertasi Program Pasca Sarjana Institut Pertanian Bogor. Fauzi A. 2010. Ekonomi Sumber Daya Perikanan Teori. Kebijakan. dan Pengelolaan. Jakarta: Gramedia Pustaka Utama

Ayunda N (2014) Efektivitas Kelembagaan Awik-Awik Pengelolaan Sumber Daya Perikanan Pantai di Kabupaten Lombok Timur (The Effectiveness of Awik-awik in Fisheries Coastal Management of East Lombok). Master thesis, Bogor Agriculture University

Ayunda N, Anna Z (2015) Evaluasi Awik-awik Pengelolaan Sumber Daya Perikanan Pantai Lombok Timur (Evaluation of local institutional, Awik-awik, in governing coastal resource utilization in East Lombok Regency). Jurnal Kebijakan Sosek KP (Soc Econom Mar Fish Pol). 5:47–57

Ayunda N, Hidayat A, Anna Z (2014) Efektivitas Kelembagaan Awik-awik dalam Mengelola Sumber daya Perikanan Pantai di Kabupaten Lombok Timur (The effectiveness of Awik-awik, the local institutional in the governing of the coastal fisheries in East Lombo*k*). Jurnal Ekonomi Pertanian Sumberdaya dan Lingkungan (J Agric Resour Environ Econom) 1(1):12–27

Bank R, Lewis A (2011) Action plan Indonesian Tuna fisheries: better management practices for moving toward sustainable and responsible fisheries. Indonesian Ministry of Marine Affairs and Fisheires and WWF pp 54

Christie P (2004) Marine protected areas as biological successes and social failures in Southeast Asia. Am Fish Soc Symp 42:155–164

Clarke RP, Yoshimoto SS, Pooley SG (1992) A bioeconomic analysis of the Northwestern Hawaiian Islands Lobster fishery. Mar Resour Econom 7:115–140

Clifton J (2003) Prospects for co-management in Indonesia's marine protected areas. Marine Policy 27:389–395

Cochrane KL (ed) (2002) A fishery manager's guidebook. Management measures and their application. FAO Fisheries Technical Paper No 424, pp 231

CSR: Creative Research System (2016) Correlation. https://www.surveysystem.com/correlation.htm. Accessed 26 Jun 2017

DJPT: Indonesia Directorate General of Capture Fisheries, (2011) Capture fisheries statistic of Indonesia 2010. Directorate General of Capture Fisheries, Jakarta, p 134

DJPT: Indonesia Directorate General of Capture Fisheries (2013) Capture fisheries statistic of Indonesia 2011. Directorate General of Capture Fisheries, Jakarta, p 134

DJPT: Indonesia Directorate General of Capture Fisheries (2015) Statistic of marine capture fisheries by fisheries management area (FMA) 2005–2014. Directorate General of Capture Fisherie, Jakarta, p 517

Eero M, Koster FW, Vinther M (2012) Why is the Eastern Baltic cod recovering? Mar Pol 36: 235–240

Eero M, Hjelm J, Behrens J, Buchmann K, Cardinale M, Casini M, Gasykov P, Holmgren N, Horbowy J, Kirkegaard KHE, Kornilovs G, Krumme U, Koster FW, Oeberst R, Plikshs M, Radtke K, Raid T, Schmidt J, Tomczak MT, Vinther M, Zimmermann C, Storr-Paulsen M (2015) Eastern Baltic cod in distress: biological change and challenges for stock assessment. ICES J Mar Sci 72:2180–2186

FAO: Food and Agriculture Organization on United Nation (1995) Code of conduct for responsible fisheries, pp 41

FAO: Food and Agriculture Organization on United Nation (2009) FAO fishery glossary. http://www.fao.org/faoterm/en. Accessed: 11 Oct 2016

FAO: Food and Agriculture Organization on United Nation (2014) Fisheries and aquaculture profile: Indonesia. http://www.fao.org/fishery/facp/IDN/en#CountrySectorSectorSocioEcoContribution. Accessed 20 Mar 2017

FAO: Food and Agriculture Organization on United Nation (2015) Voluntary guidelines for securing sustainable small-scale fisheries in the context of food security and poverty eradication. FAO, pp 20

Fauzi A, dan Anna S (2005) Pemodelan Sumber Daya Perikanan dan Kelautan untuk Analisis Kebijakan (Modeling of Marine and Fisheries Resources for Policy Analysis). PT. Gramedia Pustaka Utama, Jakarta

FLO: Fishing-Living Organization (2014) Indonesia Fair Trade Certification. http://fishing-living.org/indonesia-fair-trade-certification/#sthash.AoB2gJcc.dpbs. Accessed 18 June 2017

Gallic BL (2002) Fisheries sustainability indicators: the OECD experience. Join workshop EEA-EC DC Fisheries-DG Enviroment on Tools for measuring (integrated) fisheries policy aiming at sustainable ecosystem. Workshop held in Brussels 28–29 Oct 2002

Hidayat A (2005) Institutional analysis of coral reef management a case study of Gili Indah Village West Lombok. Indonesia. Berlin. Humboldt- Unv, Shaker Verlag

KKP: Kementrian Kelautan dan Perikanan Republik Indonesia (Ministry of marine affairs and fisheries republic of Indonesia) (2014). http://kkp.go.id/visi-misi/. Accessed 18 June 2016

KOMINFO: Kementrian Komunikasi dan Informatikan Republik Indonesia (Ministry of Communcation and Information Republic of Indonesia) (2016) https://www.kominfo.go.id/content/detail/8231/menuju-poros-maritim-dunia/0/kerja_nyata. Accessed 25 Sept 2017

Kurniawan (2015) Analisis potensi dan degradasi sumberdaya perikanan cumi-cumi (Urotheutis chinensis) Kabupaten Bangka Selatan (Analysis of potential and degradation of squids resources in Regency of South Bangka). Akuatik-Water Resour J 9(1):10–17

Mochtar Z (2016) Small-scale fisheries in Indonesia. The 2nd International Symposium on Fisheries Crime 2016. Yogyakarta. Accessed 11 Oct 2016

Najamuddin, Baso A, Arfiansyah R (2016) Bio-economic analysis of coral trout grouper fish in Spermonde Archipelago, Makasar, Indonesia. Int J Oceans Oceanogr 10(2):247–264

OANDA: Online Trading & FX for Business (2017) https://www.oanda.com/currency/average. Accessed 18 Jun 2017

OCEANA (2012) Fisheries management in the Baltic Sea: how to get on track to a sustainable future in Baltic fisheries

Triyono H (2013) Total Allowable Catch (TAC) method for Marine Fisheries Resources in Fisheries Management Area of Indonesia. Paper presented at The Science Outlook in the Preparation of Ministry of Marine Affair and Fisheries Decision. Jakarta 20–22 Mar 2013

Turner SJ, Thrush SF, Hewitt JE, Cummings VJ, Funnell G (1999) Fishing impacts and the degradation or loss of habitat structure. Fish Manag Ecol J 6:401–420

Utami B P (2015) Kebijakan Ekonomi dalam Pengelolaan Perikanan Pelagis Besar Berkelanjutan di Kabupaten Flores Timur (Econmic policy in Big Pelagic fisheries management in East Flores), Master thesis of Bogor Agriculture University

The Role of Peatlands and Their Carbon Storage Function in the Context of Climate Change

Kamila M. Harenda, Mariusz Lamentowicz, Mateusz Samson and Bogdan H. Chojnicki

Abstract Peatlands are unique habitats that are covering around 3% of the land area and they are characterized by high sensitivity to climate. These very complex ecosystems impact both water and carbon cycle at local as well as global scale. Peatlands are also valuable ecosystems due to their mitigating features in terms of floods or soil erosion and they can store and filtrate water in the landscape as well. As a result of high moisture they can also gather a big amount of carbon and this ability makes peatlands climate coolers. On the other hand a stored carbon can be released into the atmosphere due to peat moisture decrease and it accelerate the global warming processes. Beside climate changes, peatlands are under pressure that is caused by human activities like land use changes or fires. Peatlands protection and restoration can both mitigate climate changes and water balance disturbances. A review of peatlands status and feature in the context of climate changes and human-induced disturbances are presented in this paper.

Keywords Peatlands protection and restoration · Carbon storage
Climate change

K. M. Harenda (✉) · M. Samson · B. H. Chojnicki
Meteorology Department, Poznan University of Life Sciences, Piątkowska 94,
60-649 Poznań, Poland
e-mail: kamilaharenda@gmail.com

M. Lamentowicz
Department of Biogeography and Palaeoecology, Faculty of Geographical
and Geological Sciences, Adam Mickiewicz University, Ul. B. Krygowskiego 10,
61 680 Poznań, Poland

M. Lamentowicz
Laboratory of Wetland Ecology and Monitoring, Faculty of Geographical
and Geological Sciences, Adam Mickiewicz University, B. Krygowskiego 10,
PL 61 680 Poznań, Poland

T. Zielinski et al. (eds.), *Interdisciplinary Approaches for Sustainable Development Goals*, GeoPlanet: Earth and Planetary Sciences,
https://doi.org/10.1007/978-3-319-71788-3_12

1 Introduction

Water is the basic component of many processes on the Earth and its permanent presence in the landscape causes the emerging of many unique and valuable habitats such as peatlands. They are very sensitive ecosystems that arise due to organic matter accumulation and high water table (Kulczyński 1949; Whiting and Chanton 2001; Keddy 2002; Erwin 2009; Rydin and Jeglum 2013). The main factor is inundation, which stimulates peat forming (mainly by anaerobic processes).

Globally, peatland habitats are rare (Gorham 1991; Keddy 2002) since they cover approximately 3% of the land area on the Earth (Clymo et al. 1998; Blodau 2002; Rydin and Jeglum 2013) and store one-third of global soil carbon (Gorham 1991; Lappalainen 1996; Page et al. 2011; Rydin and Jeglum 2013). Most peatlands are located in the northern hemisphere (around 80% in the boreal and subarctic zones), 10% can be found in the tropics and Southeast Asia and another 10% are located in the temperate zone (Yu et al. 2010, Frolking et al. 2011, Tobolski 2012). Figure 1 shows peatlands distribution on the Earth (Main Report 2007).

The classification of peatlands can be very complex, yet the simplest one consists of four types of peatlands (Keddy 2002):

- Swamp—dominated by trees with roots in hydric soils (not in peat), like tropical mangrove swamps.
- Marsh—dominated by herbaceous plants emerging through the water and rooted in hydric soils (not in peat), like reed beds around the Baltic Sea.
- Bog—dominated by *Sphagnum* moss, shrubs, sedges or evergreen trees with roots in deep peat, like floating bogs covering shores of lakes in temperate and boreal regions.
- Fen—dominated by sedges and grasses rooted in the peat and considerable water movement through the peat is noticed, like the extensive peatlands in northern Canada.

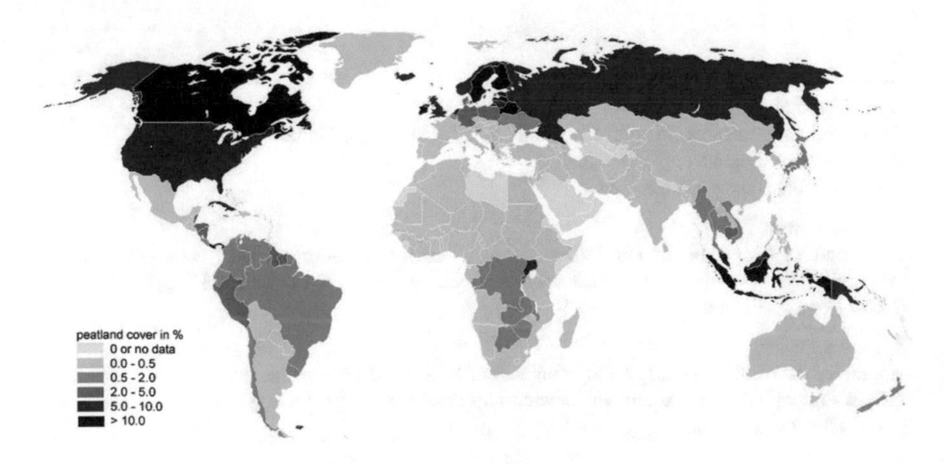

Fig. 1 Percentage peatlands distribution per country (Main Report 2007)

Peatland's ecological functions can have direct and/or indirect effects. The first group of functions is related to water, including: storage, filtration and supply. They also mitigate floods and prevent erosion. They stabilize the macroclimate and play an ecological function. Due to their ability to retain water, peatlands often prevent drought (Tobolski 2012). Indirect functions include nutrient retention, carbon storage, and sediment retention. Peatlands provide water to other ecosystems and they are rich in biological diversity (genetic reservoirs of many organisms). Their unique features create habitats suitable for many endemic and/or endangered species of plants and animals and they can form ecological corridors enabling migration (Mäkilä and Saarnisto 2008). Peatlands can be considered as sources of humus in the landscape (Keddy 2002; Mäkilä and Saarnisto 2008). Peatlands can be also considered as an archive of past climates due to their sensitivity to weather and capability of long term conservation of produced biomass (Clymo et al. 1998). The carbon accumulation process in the peat appears when the rate of organic matter decomposition is lower than the amount of primary production of the ecosystem. In their natural state, peatlands are usually carbon sinks (Turunen et al. 2002; Mäkilä and Saarnisto 2008) and this feature makes these ecosystems very important elements of the environment in the context of climate change, since the absorbed CO_2 is one of the major greenhouse gases. Furthemore, peatlands possess unexplored diversity of protists playing important role in this ecosystem (Marcisz et al. 2014; Geisen et al. 2017; Mulot et al. 2017) They contribute to carbon fixing, however this issue still needs scientific investigation (Jassey et al. 2015, 2016).

Analyses of modern and palaeoenvironmental data identify peatlands as an element of nature that is predicted to affect future climate. Therefore, further research, protection and restoration of these vulnerable ecosystems should be a priority in the context of global warming (Dise 2010; De Jong et al. 2010; Lamentowicz et al. 2016). The maintenance of those peatlands' functions requires the appropriate management and conservation. This paper contains the overview of peatland's function in the context of future environmental changes and the interactions between the climate and these valuable ecosystems.

2 Carbon Dynamics in Peatlands

Despite the fact that peatlands cover relatively small area of the land, their role in the global carbon cycle is not insignificant (Frolking et al. 1998; Moore et al. 1998; Yu et al. 2011; Charman et al. 2013; Loisel et al. 2014). There are several processes in the peatlands where carbon is involved, such as CO_2 the exchange with the atmosphere, the emission of CH_4, the production and export of dissolved organic carbon (DOC) and others (Moore et al. 1998). Furthermore, peatlands are very important stores of temporarily sequestered carbon (Clymo et al. 1998; Ilnicki and Iwaniszyniec 2002; Rydin and Jeglum 2013).

The carbon balance of peatlands is basically determined by two processes, photosynthesis and respiration. There are several factors that determine the process of photosynthesis including solar radiation, CO_2, water, air temperature and nutrients (Maćkowiak and Michalak 2008). The carbon compounds enter into peatland biomass through this process. Further transformation of this biomass is deeply dependent on hydrology. High humidity of the soil leads to anoxic conditions that slow down the decomposition processes of organic matter. The peat is formed as a result of those processes - production (photosynthesis) and decomposition (respiration) and a clear, negative correlation between those opposite processes is observed (Clymo 1984; Mäkilä and Saarnisto 2008). Decomposition is the second controlling factor of carbon accumulation. This makes the peatlands more sensitive to climatic changes than to plant composition changes (Ise et al. 2008). In general, it was found that the lowest carbon accumulation was observed at the sites with the lowest water table e.g. forested and sparsely forested mires while the highest sequestration rate was observed in young *Sphagnum* peatland in coastal regions. These costal peatlands are good examples of high carbon accumulation rate since the *Sphagnum* production is mainly controlled by the moisture that is transported from the sea (Mäkilä and Saarnisto 2008). Under the present climate conditions, differences between the carbon accumulation ability of different types of peatlands are noticeable. The boreal raised peatlands are currently a carbon sink (Turunen et al. 2002), whereas the northern circumpolar wet aapa-mires are source of greenhouse gases and their carbon accumulation rate is lower than most southern raised bogs (Mäkilä and Saarnisto 2008).

In addition, part of pristine areas, mainly tropical and permafrost, may no longer accumulate the peat because of recent climate changes (Main Report 2007).

Peatlands are second only to oceans natural stock of carbon and they can contain twice as much carbon than the entire world's biomass (Rydin and Jeglum 2013). This is the effect of high carbon concentration in the peat (approx. 50%). Peatland ecosystems are characterized by low productivity but long-term storage. However, the length of the storage time depends on the wetness of the substrate (Strack 2008).

In general, the absorption of CO_2 from the atmosphere over 12,000 years allowed for the large quantity storage of carbon in the peat (Yu et al. 2010). The global estimation of the peatland's carbon storage is problematic mainly due to the difficulties of precisely defining the area of peatlands and the assessment of the thickness of peat layer. Recently, large areas of peatlands were also discovered inter alia in Peruvian Pastaza-Marañón (Draper et al. 2014) and the Congo Basin (Dargie et al. 2017). These findings determined the increase of the estimation of tropical peatland carbon stock by 36%, to 104.7 GtC (Dargie et al. 2017). Currently the global amount of carbon that is stored in the peat is estimated to exceed 600 GtC and the sizes of northern, southern and tropical stocks are approximately 550 GtC, 15 GtC and 100 GtC, respectively (Yu et al. 2010; Köchy et al. 2015; Abrams 2016; Kleinen et al. 2016).

The amount of this resource is inter alia the result of ice ages when the peatlands' role in the global carbon cycle was clearly noticeable. The CO_2 concentration in the atmosphere during this period dropped by approx. 100 ppm and one of the explanations of this fact was high water saturation of the high-latitude soils that slowed down the rate of organic matter decomposition (Zech et al. 2011). This factor has the biggest importance in the carbon accumulation process in the soil.

When the climate was getting warmer (after the glacial period), the permafrost started to disappear and the result of this process was the emission of trapped soil carbon both in gas (CO_2 and CH_4) and dissolved in water (DOC-dissolved organic carbon and POC-particulate organic carbon) forms (Limpens et al. 2008; Billett et al. 2010). Simultaneously, higher concentrations of CO_2 in the atmosphere and higher air temperature during summer improved the effectiveness of the photosynthetic process that induced a rapid growth of vegetation. The high latitude areas that were released from the ice cover became a suitable environment for new peatlands to form. Due to continuous spatial expanding of peatlands during the early Holocene period years ago, a noticeable drop in CO_2 concentration was found (Zech et al. 2011). The linkage between thawing permafrost and expanding forest was also found. When the ice was melting down and peatlands of Arctic North America and Eurasia were releasing the greenhouse gases, the expansion of forests northward into tundra compensated for this emission (Mäkilä and Saarnisto 2008).

It was also found that local environmental factors and natural succession of peatlands may attenuate the regional relationships between climatic factors and observed stratigraphical and hydrological changes of peatlands in Finland (Mäkilä and Saarnisto 2008).

3 Hydrology

The minimum ambient temperature provides the conditions for peatland growth if the proper amount of water is available (Yu et al. 2009). Therefore, water, in the context of peatlands, is considered a first-order factor determining their occurrence, growth and development (Keddy 2002). The water balance of peatland basically consists of precipitation, evapotranspiration, runoff and retention and the climate solely regulates all presented elements (Charman and Mäkilä 2003; Mäkilä and Saarnisto 2008). In the temperate climate conditions, water is the most important ecological factor that determines two types of peatlands: rheotrophic mires (fens) that are fed directly by both rain- and ground-water flow, and ombrotrophic mires (raised bogs) that rely on water input only in the form of rainfall (Moore et al. 1998; Tobolski 2012). In the case of ombrotrophic raised bogs it is impossible to detect the impact of surface runoff and groundwater so the only reasonable components of the water balance are precipitation and evapotranspiration (Charman and Mäkilä 2003).

Peat growth is related to the amount of precipitation and temperature, thus the changes in the hydrology can be studied by analyzing the changes of the peat accumulation rate in the peat profile (Yu et al. 2009). Paleoecological studies show that more peat was accumulated during wet periods in Holocene and it had an effect of the higher growth of existing peatlands as well as their proliferation in the landscape (Yu et al. 2003). Conversely, during the Holocene's dry periods the rate of carbon accumulation dropped to about half of the present level. These changes caused by the precipitation reduction were found at both the northern and tropical peatlands (Strack 2008). These findings make the peatlands very large natural carbon stock highly vulnerable and susceptible to any changes of hydrological cycles (Erwin 2009; Belyea and Malmer 2004; Clymo et al. 1998; Dise et al. 2011).

The complexity of peatland's functions and its high vulnerability to many factors impede the detailed assessment of the future dynamics of these ecosystems. Nowadays, palaeoecological and observational studies strongly suggest that peatlands will be affected mainly by climate change and human determined disturbances (Turetsky et al. 2002; Mauquoy and Yeloff 2007; Yu 2007; Lamentowicz et al. 2008).

4 Climate Change

Climate change can be considered as both natural (Erwin 2009) and anthropogenic (IPCC 2013) and it has been associated with a range of weather-related disasters, including droughts, windstorms, ice storms and wildfires (Canadian University of Waterloo report). Moreover, some climate changes will certainly have impact on peatlands' hydrology but other changes may cause an increase in local and regional temperature, alter evapotranspiration, biogeochemistry, amounts of suspended sediment loadings, fire, oxidation of organic sediments and the physical effects of wave energy (Erwin 2009; Burkett and Kusler 2002). Global air temperatures increased by about 0.7 °C during the period of 1906–2005 (IPCC 2013). Furthermore, on the basis of general climate models scientists predict that the temperature will increase by additional 2–8 °C by the end of the 21st century, depending on the region (Christensen et al. 2007; IPCC 2013; Erwin 2009). IPCC also predicts changes in precipitation that consider both extreme rainfall periods and periods of drought (IPCC 2013).

Recently, global estimation of carbon loss from the upper soil horizon due to 1 °C increase varies from 30 to 203 GtC by 2050 (Crowther et al. 2016). This temperature could induce a global increase in heterotrophic respiration of 0.038–0.100 GtC per year (Dorrepaal et al. 2009). Moreover, the amount of global CH_4 emission from the peatlands is estimated at approximately 0.123 GtC per year. Due to the fact that at high latitudes the air temperature will increase to a greater extent in the next century, the emissions of CH_4 from the northern areas can increase disproportionately (Bridgham et al. 2013). It is believed we can expect even bigger loss of carbon in the form of methane in the future.

5 Fires

The role of peatland fires in the carbon cycle deserves increased scientific attention (Gorham 1991; Strack 2008). Recently, scientists have observed the climate warming causing longer and more severe drought periods that in the near future may result in higher fire activity (Higuera 2015; Kettridge et al. 2015). These heat waves may amplify the fire activity in peatlands and lead to a release of huge amounts of carbon dioxide into the atmosphere (Turetsky et al. 2011, 2015). The enormous load of carbon that has remained in peat over millennia will be converted into the atmospheric part of the global carbon cycle. The peat fires can be also the source of non-CO_2 greenhouse gases in the atmosphere. For instance, anoxic conditions present at peatlands result from the fact that fires of these ecosystems can emit around ten times more CH_4 per unit of combusted biomass than fires of savannas (Yokelson et al. 1997; Andreae and Merlet 2001; Christian et al. 2003). These increased GHG emissions may cause the intensification of the greenhouse effect and lead to rising temperatures which intensify the frequency of climate determined fires. This positive feedback is amplified by more frequent boreal forests fires (Randerson et al. 2006; Flanningan et al. 2001). Moreover, peatland fires can smolder for months due to the thick peat layer and pose a threat of further fires in the vicinity of peatlands (Benscoter et al. 2015). These conditions can be observed, especially in autumn when the peat moisture is relatively low (The Guardian 2016). The peat destroyed by the fire is the source of mineral compounds in the environment (Strack 2008). Thus, this temporary heavy fertilization results in the withdrawal of plants not adapted to such conditions (Kuhry 1994; Sillasoo et al. 2011). The presence of permafrost in the peat profiles of the northern peatlands effectively limits the depth of peat burning but the projections of its thawing reduces the fire resistance of these ecosystems (Natural Resources Canada 2016).

Recently, the tropical peatlands destruction became an important environmental issue. These usually pristine ecosystems were systematically destroyed by drainage and burning and these activities altered their water and carbon retention capabilities. As a result, the peatlands' carbon balance was shifting from a sink to a source (Wösten et al. 1997; Page et al. 2002; Canadell et al. 2007; Rieley et al. 1996). Page et al. 2002 estimate that due to big fires in 1997 the dried peatlands of Indonesia released 0.8–2.6 GtC both from peat and vegetation, which represents around 13–40% of the mean annual global carbon emissions from burning of fossil fuels in the world. As a result, in 1997–2009 the contribution of peatland fires to total global emissions increased from 4% to 5% (Werf et al. 2010). Tropical peatlands fire with vegetation changes have a meaningful impact on the carbon cycle, the atmosphere, the ecosystem services and they cause wide-ranging social and economic impacts too. Recently, fires at extensive tropical peatlands have become more regular and the biggest ones were linked with the El Niño phase of ENSO that causes long drought periods (Page et al. 2009a). Rapid land use changes and climatic variability led to an increase of fire frequency in recent decades. In addition, highly variable, from year to year, carbon emission from this source will not be balanced by peatland's regrowth that follow the fire (Werf et al. 2010).

Peatlands and their carbon stock can be also affected by fires of the vegetation in the surrounding areas. Fires causing local or regional deforestation affect hydrology of peatlands of different size—from small kettle-hole peatlands to extensive raised bogs. It was shown for example that large fires of the Noteć forest in W Poland caused not only hydrological changes (wet shift/aqualysis) but also triggered floating mats development in some wetlands (Lamentowicz et al. 2015). Furthermore, a high-resolution study by Marcisz et al. (2015) revealed how important might be an indirect impact of fires in a peatland catchment that was also inferred recently from a peat core in S Poland (Kajukalo et al. 2016).

6 Peatlands Reduction

The disappearance of peatlands from the landscape had various spatial and temporal intensity but it was always caused by the economic and/or vital needs of human. Peatlands are potential farming areas thus they have been disturbed or completely degraded by human activities over the last two centuries. Those transformations were initially realized in Europe. For instance, 90% of peatlands in Switzerland disappeared mainly due to conversion to agricultural, garden and vegetable crops (Tobolski 2012; Lamentowicz et al. 2007).

This land-use change was mainly done by drainage that causes water table level instabilities. This factor led to deeper aeration of acrotelm that resulted in abrupt change - the replacement of plant communities and substantial transformation of the peatland ecosystem (Milecka et al. 2016). The peat extraction for fuel and industrial purposes is the second reason of peatland reduction (Strack 2008). There are large quantities of peat harvested for electricity production in Finland, Ireland and Sweden (World Energy Council 2013). Peatlands also disappear from the landscape due to transformation in order to change the landscape retention e.g. river valley flooding purposes or by irresponsible destruction e.g. the solid minerals, petroleum and natural gas mining (Tobolski 2012). A good example of the negative impact of mining activity was recently observed in the destruction of peatlands in Canada. The preparation of these areas for bituminous sand mines caused the release of around 11.4–4.73 million tons of stored carbon (Rooney et al. 2011).

Recently, socio-economic changes in Southern Asia caused enormous damage of tropical peatlands (Page et al. 2009b). The average peat thickness of tropical peatlands is around 5 m (maximum depths are over 20 m) while the mean thickness of peat observed at higher latitudes is lower than in Finland which is around 1.2 m (Page et al. 2011, Tropical peatlands-University of Helsinki). A place of special importance in the context of climate protection is South-eastern Asia. Peatlands in this region are significant terrestrial carbon storage, both in aboveground biomass and thick deposits of peat (Page et al. 2009a; Hamada et al. 2013).

In addition, tropical peatlands have far greater ability to accumulate carbon. Although they represent only 10% of the world's peatlands, they are responsible for up to 37% of this potential. Tropical peatlands also contain 10 times more carbon per hectare than ecosystem on mineral soils, whereas subpolar and boreal zone peatlands contain 3.5 and 7 times more carbon, respectively (Strack 2008).Tropical peatlands are an important component of the global terrestrial carbon resource because they store around 20% of carbon of all peatlands in the world in both their aboveground biomass and underlying thick deposits of peat (Page et al. 2011, Tropical peatlands-University of Helsinki).

Nowadays most of these ecosystems are systematically destroyed mainly by preparing the soil for palm oil plantations. These land use changes are realized by deforestation, fire and drainage because dry peat is much more prone to wildfires (Werf et al. 2010). This unprecedented scale of destruction converts the peatlands into a big source of atmospheric CO_2 (Moore et al. 2013; Page et al. 2011; Hooijer et al. 2010; Yale Environment 2017). Due to land use changes in 1990–2015, the carbon reservoir in tropical southeastern peatlands has decreased by about 2.5 GtC which corresponds to several hundred or even several thousand years of carbon accumulation in peatlands (Miettinen et al. 2017).

Moreover, the rate of peat exploitation is more than twice as fast as in its formation and it causes the decrease of peat by approximately 20 km^3 (Main Report 2007). These disturbances also modify the peat carbon dynamics due to the fact that since 1990, southeast Asian disturbed peatlands caused a 32% increase in fluvial organic carbon flux. This increase is more than half of the entire annual fluvial organic carbon flux from all European peatlands. Moreover, altering the structure of peatlands causes higher DOC release from deep peat layers that were formed thousands of years ago (Moore et al. 2013).

Total peat destruction would be equivalent to 100 years of coal burning at current rates. Additionally, the part of peatlands (including tropical and northern permafrost ones) doesn't accumulate peat anymore due to global warming. Consequently, the actual rate of carbon accumulation by all peatlands over the world does not exceed 0.1 GtC per year (Main Report 2007).

7 Peatlands Protection and Restoration

The enormous scale of destruction of these ecologically valuable elements of nature raised concerns and questions about the fate of these carbon depots and the interaction between the climate and peatlands in the future (Joosten et al. 2017; Bonn et al. 2016).

There are two modes of peatland's protection, passive and active (Tobolski 2012). The passive protection consists of total or maximum exclusion of human activities with the exception of preventing adverse changes introduced into the ecosystems by human activity. Usually it is realized by the introduction of different conservation strategies. The active protection is applied at the objects where human intervention is necessary. It leads to the preservation of peatlands along with rare and peat-forming peatland species. For example, mowing is introduced in order to stop vegetation succession on meadows (Hedberg et al. 2013; Kotowski et al. 2016) or even removal of the humified peat from degraded fens (Klimkowska et al. 2010). Moreover, large scale, long-term projects were utilized to restore raised bogs in N Poland (Herbichowa 2007).

Peatland protection is also carried out on different spatial scales. For example, the Ramsar convention that is the global intergovernmental treaty (adopted in 1971) provided the framework for conservation and wise use of peatlands (Erwin 2009; Matthews 1993). It was one of the first modern legal instruments to conserve natural resources on a global scale. It links the countries and restrains from unreasonable exploitation of natural ecosystems. This convention's standards of wise use principles of management and protection were adopted in the international arrangements and the national laws (Matthews 1993). The Natura 2000 was a network of nature protection areas that was adopted in 1992 for the protection of the most seriously threatened habitats in the European Union territory. Peatlands are on the list of the habitats that are protected by this form of conservation (Natura 2000). Special attention to peatlands is also espoused by various countries. For example, the Swiss federal constitution expressly states the importance of peatland protection (Tobolski 2012), while in Poland the statutory form of peatlands protection is peatland reserve and they are also protected in the framework of national parks, ecological grasslands or documentation sites (Nature protection).

The peatlands that are already disturbed or destroyed can be reestablished with restoration processes. Due to the fact that water is the most important factor determining of the existence of peatlands, rewetting is most often the solution for restoration of perturbed objects (Tuittila et al. 2000; Zerbe et al. 2013). However, increased water level does not ensure successful restoration since high ground moisture of can be insufficient for the restoration of peatlands.

Regardless of doubts about its effectiveness, rewetting can be accomplished in many practical ways, e.g. construction of small retention devices to stop the water outflow with ditch drainage system (Glińska-Lewczuk et al. 2014; Bonn et al. 2016). Raising water levels to the soil surface or even above to maintain anaerobic soil conditions results in environments that are good enough for rapid succession towards closed peatland vegetation (Tuittila et al. 2000; Zerbe et al. 2013). Water management gives promising results at many locations. Zerbe et al. (2013) found that ten years after rewetting, manipulated peatlands in North East Germany formed mosaic of vegetation types with the highest potential for peat formation. The interesting case of positive impact of human activity that positively influences the peatlands can be found in maritime areas of Canada. The stretches of dikes that protect agricultural land, infrastructure, homes and communities additionally inhibit salt marshes from naturally shifting with the level of the sea and reducing destructive wave action (Erwin 2009).

8 Discussion and Summary

Peatlands are very important terrestrial ecosystems due to their uniqueness, vulnerability and importance in the global water and carbon cycles as well as climate forming. The arising and development of these ecosystems was always determined by complex of thermal and hydrological conditions and due to long-term influence these factors peatlands are the second, after the ocean, stock of carbon in the world.

Peatlands interacted with atmosphere and cooling climate during the ice age by storing greenhouse gases (GHG) related carbon. Some Holocene climate studies suggest that peatlands are supposed to have cooled down the climate recently and only anthropogenic emissions have prevented the initiation of the Ice Age (Rudimann 2003).

Nowadays, the human activity and climate changes are expected to be the most important factors determining the peatland's carbon sequestration (Charman et al. 2013; Loisel et al. 2014). Additionally, the fires induced by both mentioned above factors will become the serious threat for these ecosystems. This set of simultaneously acting factors affect both the hydrology, land use, temperature (direct and indirect way) (Ferretti et al. 2005) and the intensification of extreme weather events that also will have a destructive impact on every kind of ecosystem in the world. There are accompanying effects that usually play role at local scale such as altered hydrology, base flow shifts, decreased water resource, increased heat stress, soil erosion, increased flood and risk of fires etc. (Erwin 2009). All the factors described above affect the stability of peatlands carbon stocks and finally lead to bigger emission of carbon and warming up the atmosphere. These changes in the functioning of peatlands transform many of them from climate 'coolers' (carbon net sinks) to the 'heaters' (carbon net sources) and it can be considered as very serious threat due to their possible positive feedback with climate. Additionally, the CO_2 uptake ability of peatlands will not be able even to compensate the anthropogenic GHG emissions (Mäkilä and Saarnisto 2008; Petrescu et al. 2015).

The study of carbon exchange between the peatlands and the atmosphere is necessary also because of high uncertainty of the estimation of CO_2 concentration rise as a result of climate warming. The recent projections of CO_2 increase as an effect of 1 °C temperature increase ranges from 1.7 to 60 ppm CO_2 (Frank et al. 2010; Cox and Jones 2008) and many climate change models do not include peatlands or coral reefs ($CaCO_3$) in the estimation of the future CO_2 concentration in the atmosphere.

Peatlands' function as landscape water buffer that through the quick absorption of precipitation can reduce the effects of floods are sometime neglected in the short-term economy. Values of these ecosystems can be described by the following example. The Canadian peatlands help to save around 3 million at rural site and even 50 million at urban site and it was also estimated that dehydration of peatlands under Canadian conditions will increase the cost of floods damages by 29% and 38% in rural and urban areas, respectively (Canadian University of Waterloo report). The forest succession in the context of peatlands was also found and it is an element that mitigated the emission from the peatlands during the Holocene melting period however recently this shift of plant cover can be limited by the wood harvesting within the regular forest management. The forest surrounding the peatland can be an effective barrier that prevents peatlands of fertile water impact from the surrounding fertilized fields; therefore, the forests (especially dominated by Pinus sylvestris) might affect peatlands' trophic state in long-term context (Lamentowicz et al. 2007; Milecka et al. 2016).

There is no doubt that the global changes are unavoidable and because of that fact the suitable management and protection strategy need to be implemented to ensure the future peatlands sustainability. Since, mentioned before, the important role of these ecosystems in the global carbon cycle the comprehensive understanding of the functioning of peatlands under climate change as well as human impact that is critical for adequate protection's strategy of peatlands. Additionally, the protection of each object requires the basic knowledge about the local history, ecology and specific character of the habitat which requires its precise analysis (Erwin 2009; Page et al. 2011; Tobolski 2012). Moreover, the choice of protection method is critical because of complex character of the influence of each implemented change.

Activities that will allow the effective protection of peatlands should include good protection policy of peatlands that roots from general approaches expressed in global/continental scale conventions and ends up with locally related protection goals and tasks. Additional efforts necessary for successful protection are to increase the public knowledge related to protected objects. The education is fundamental to gain the success in peatland protection, especially on local scale. The effectiveness of maintaining the peatlands as the carbon sinks is directly related to the ecological awareness of the local communities. Another tool that will support the introduction of reasonable protection management is the appropriate monitoring program (Słowińska et al. 2010) that provides the information both long and short-term changes of these vulnerable ecosystems. Such system provides insights to the potential ecological consequences of the changes, supporting the decision makers to determine the management practices that should be implemented. It helps also the understanding the range of current variability in some parameters and detecting desirable and undesirable changes in time within peatland areas and adjacent ecosystems (Erwin 2009). These activities will allow to introduce adequate solutions to mitigate the effects of climate change and reduce or eliminate the human impact on these valuable ecosystems.

Recently, beside protection, people have tried also to restore disturbed or completely destroyed areas of the peatlands in the world (e.g. Werf et al. 2010). Appropriate techniques are selected on the basis of local conditions and they are related to water since this factor determines the existence of peatlands (Erwin 2009). Such actions have been taken inter alia in Germany and Canada and in these cases it was successful (Zerbe et al. 2013). However, one must be aware that restoration might not always be effective because past disturbances were too extensive.

Techniques that are often used concern rewetting and building the small retention devices in order to water retention in the peatlands or rising the water level. This promotes the biological diversity and strongly moistened conditions cause also the spreading of peatland vegetation that definitely has a positive effect on the peatlands regeneration. A certain threat to the peatlands restoration may be a fact of increased global water demand since it increased more than triple since 1950 and in the future it is going to be doubled again by 2035 (Postel 1997). But, on the other hand, peatlands as a store of water in the context of floods can offset the increasing water demand. Summarizing this study, peatlands need to be protected and restored as an element of sustainable development of global civilization since it is very important in the context of mitigating the global warming effect.

Acknowledgements This work was supported by Swiss Contribution to the enlarged European Union (No. PSPB-013/2010) and the National Science Centre, Poland (grant No. NN306060940 and 2015/17/B/ST10/01,656) and by the Polish-Norwegian Research Programme, project ID: 203258, contract No. Pol-Nor/203258/31/2013.

References

Abrams JF (2016) Impacts of Indonesian peatland degradation on the coastal ecosystems and the global carbon cycle. Dissertation, Jacobs University, Bremen

Andreae MO, Merlet P (2001) Emission of trace gases and aerosols from biomass burning. Glob Biogeochem Cycl 15:955–966

Belyea LR, Malmer N (2004) Carbon sequestration in peatland: patterns and mechanisms of response to climate change. Glob Change Biol 10:1043–1051

Benscoter BW, Greenacre D, Turetsky MR (2015) Wildfire as a key determinant of peatland microtopography. Can J For Res 45(8):1133–1137

Billett MF, Charman DJ, Clark JM, Evans CD, Evans MG, Ostle NJ, Worrall F, Burden A, Dinsmore KJ, Jones T, McNamara NP, Parry L, Rowson JG, Rose R (2010) Carbon balance of UK peatlands: current state of knowledge and future research challenges. Clim Res 45:13–29

Blodau C (2002) Carbon cycling in peatlands—A review of processes and controls. Env Rev 10(2):111–134

Bonn A, Allott T, Joosten H, Evans M, Stoneman R (2016) Peatland restoration and ecosystem services: science, policy and practice. Cambridge University Press, UK

Bridgham SD, Cadillo-Quiroz H, Keller JK, Zhuang Q (2013) Methane emissions from wetlands: biogeochemical, microbial, and modeling perspectives from local to global scales. Glob Change Biol 19:1325–1346

Burkett V, Kusler J (2002) Climate change: potential impacts and interactions in wetlands of the United States. JAWRA J Am Water Res Assoc 36(2):313–320

Canadell JG, Le Quéré C, Raupach MR, Field CB, Buitenhuis ET, Ciais P, Conway TJ, Gillett NP, Houghton RA, Marland G (2007) Contributions to accelerating atmospheric CO_2 growth from economic activity, carbon intensity, and efficiency of natural sinks. P Natl Acad Sci USA 104:18866–18870

Canadian University of Waterloo report (2017) http://www.intactcentreclimateadaptation.ca/wp-content/uploads/2017/07/When-the-Big-Storms-Hit.pdf. Accessed on 30 Jul 2017

Charman D, Mäkilä M (2003) Climate reconstruction from peatlands. PAGES Newsletter 11:15–17

Charman DJ, Beilman DW, Blaauw M, Booth RK, Brewer S, Chambers FM, Christen JA, Gallego-Sala A, Harrison SP, Hughes PDM, Jackson ST, Korhola A, Mauquoy D, Mitchell FJG, Prentice IC, van der Linden M, De Vleeschouwer F, Yu ZC, Alm J, Bauer IE, Corish YMC, Garneau M, Hohl V, Huang Y, Karofeld E, Le Roux G, Loisel J, Moschen R, Nichols JE, Nieminen TM, MacDonald GM, Phadtare NR, Rausch N, Sillasoo Ü, Swindles GT, Tuittila ES, Ukonmaanaho L, Väliranta M, van Bellen S, van Geel B, Vitt DH, Zhao Y (2013) Climate-related changes in peatland carbon accumulation during the last millennium. Biogeosciences 10(2):929–944

Christensen JH, Hewitson B, Busuioc A, Chen A, Gao X, Held R, Jones R, Kolli RK, Kwon WK, Laprise R, Magaña Rueda V, Mearns L, Menéndez CG, Räisänen J, Rinke A, Sarr A, Whetton P, Arrit R, Benestad R, Beniston M, Bromwich D, Caya D, Comiso J, de Elia R, Dethloff K (2007) Near-term Climate Change: projections and Predictability. In: Solomon S, Qin D, Manning M, Chen Z, Marquis M, Averyt KB, Tignor M, Miller HL (eds) Climate Change 2007: The Physical Science Basis. Contribution of Working Group I to the Fourth Assessment Report of the Intergovernmental Panel on Climate Change, Cambridge University Press, UK, p 847–940

Christian TJ, Kleiss B, Yokelson RJ, Holzinger R, Crutzen PJ, Hao WM, Saharjo BH, Ward DE (2003) Comprehensive laboratory measurements of biomass-burning emissions: 1. Emissions from Indonesian, African, and other fuels. J Geophys Res. https://doi.org/10.1029/2003JD003704

Clymo RS (1984) The limits to peat bog growth. Phil Trans Royal Soc London B 303:605–654

Clymo RS, Turunen J, Tolonen K (1998) Carbon Accumulation in Peatland. Oikos 81(2): 368–388. https://doi.org/10.2307/3547057

Cox P, Jones C (2008) Illuminating the Modern Dance of Climate and CO_2. Science 321: 1642–1644

Crowther TW, Todd-Brown KEO, Rowe CW, Wieder WR, Carey JC, Machmuller MB, Snoek BL, Fang S, Zhou G, Allison SD, Blair JM, Bridgham SD, Burton AJ, Carrillo Y, Reich PB, Clark JS, Classen AT, Dijkstra FA, Elberling B, Emmett BA, Estiarte M, Frey SD, Guo J, Harte J, Jiang L, Johnson BR, Kröel-Dulay G, Larsen KS, Laudon H, Lavallee JM, Luo Y, Lupascu M, Ma LN, Marhan S, Michelsen A, Mohan J, Niu S, Pendall E, Peñuelas J, Pfeifer-Meister L, Poll C, Reinsch S, Reynolds LL, Schmidt IK, Sistla S, Sokol NW, Templer PH, Treseder KK, Welker JM, Bradford MA (2016) Quantifying global soil carbon losses in response to warming. Nature. https://doi.org/10.1038/nature20150

Dargie GC, Lewis SL, Lawson IT, Mitchard ETA, Page SE, Bocko YE, Ifo SA (2017) Age extent and carbon storage of the central Congo Basin peatland complex. Nature. https://doi.org/10.1038/nature21048

De Jong R, Blaauw M, Chambers FM, Christensen TR, De Vleeschouwer F, Finsinger W, Fronzek S, Johansson M, Kokfelt U, Lamentowicz M, LeRoux G, Mitchell EAD, Mauquoy D, Nichols JE, Samaritani E, van Geel B (2010) Climate and Peatlands. In: Dodson J (ed) Changing Climates, Earth Systems and Society. Series: International Year of Planet Earth. Springer, Heidelberg, p 85–121

Dise NB (2010) Peatland response to global change. Science 326:810–811

Dise NB, Narasinha JS, Weishampel P, Verma SB, Verry ES, Gorham E, Crill PM, Harriss RC, Kelley CA, Yavitt JB, Smemo KA, Kolka RK, Smith K, Kim J, Clement RJ, Arkebauer TJ, Bartlett KB, Billesbach DP, Bridgham SD, Elling AE, Flebbe PA, King JY, Martens CS, Sebacher DI, Williams CJ, Wieder RK (2011) Carbon emissions from peatlands. Peatland Biogeochemistry and Watershed Hydrology at the Marcell Experimental Forest. CRC Press, USA, p 297–347

Dorrepaal E, Toet S, Van logtestijn RSP, Swart E, Van De Weg MJ, Callaghan TV, Aerts R (2009) Carbon respiration from subsurface peat accelerated by climate warming in the subarctic. Nature 460:616–619

Draper FC, Roucoux KH, Lawson IT, Mitchard ETA, Coronado ENH, Lähteenoja O, Montenegro LT, Sandoval LV, Zaráte R, Baker TR (2014) The distribution and amount of

carbon in the largest peatland complex in Amazonia. Environmental Research Letters. https://doi.org/10.1088/1748-9326/9/12/124017
Erwin KL (2009) Peatlands and global climate change: the role of peatland restoration in a changing world. Wetlands Ecol Manag 17:71–84
Ferretti DF, Miller JB, White JWC, Etheridge DM, Lassey KR, Lowe DC, MacFarling Meure CM, Dreier MF, Trudinger CM, van Ommen TD, Langenfelds RL (2005) Unexpected changes to the global methane budget over the past 2000 years. Science. https://doi.org/10.1126/science.1115193
Flannigan M, Campbell I, Wotton M, Carcaillet C, Richard P, Bergeron Y (2001) Future fire in Canada's boreal forest: paleoecology results and general circulation model - regional climate model simulations. Can J For Res 31:854–864
Frank DC, Esper J, Raible CC, Buntgen U, Trouet V, Stocker B, Joos F (2010) Ensemble reconstruction constraints on the global carbon cycle sensitivity to climate. Nature. https://doi.org/10.1038/nature08769
Frolking SE, Bubier JL, Moore TR, Ball T, Bellisario LM, Bhardwaj A, Carroll P, Crill PM, Lafleur PM, McCaughey JH, Roulet NT, Suyker AE, Verma SB, Waddington JM, Whiting GJ (1998) Relationship between ecosystem productivity and photosynthetically active radiation for northern peatlands. Glob Biogeochem Cycl 12(1):115–126
Frolking S, Talbot J, Jones MC, Treat CC, Kauffman JB, Tuittila ES, Roulet N (2011) Peatlands in the Earth's 21st century climate system. Env Rev 19:371–396. https://doi.org/10.1139/a11-014
Geisen S, Mitchell EAD, Wilkinson DM, Adl S, Bonkowski M, Brown MW, Fiore-Donno AM, Heger TJ, Jassey VEJ, Krashevska V, Lahr DJG, Marcisz K, Mulot M, Payne R, Singer D, Anderson OR, Charman DJ, Ekelund F, Griffiths BS, Rønn R, Smirnov A, Bass D, Belbahri L, Berney C, Blandenier Q, Chatzinotas A, Clarholm M, Dunthorn M, Feest A, Fernández LD, Foissner W, Fournier B, Gentekaki E, Hájek M, Helder J, Jousset A, Koller R, Kumar S, La Terza A, Lamentowicz M, Mazei Y, Santos SS, Seppey CVW, Spiegel FW, Walochnik J, Winding A, Lara E (2017) Soil protistology rebooted: 30 fundamental questions to start with. Soil Biol Biochem 111:94–103
Glińska-Lewczuk K, Burandt P, Łaźniewska I, Łaźniewski J, Menderski S, Pisarek W (2014) Ochrona i renaturyzacja torfowisk wysokich w rezerwatach Gązwa, Zielony Mechacz i Sołtysek w północno-wschodniej Polsce. Wydawnictwo Polskiego Towarzystwa Ochrony Ptaków, Białowieża
Gorham E (1991) Northern peatlands: role in the carbon cycle and probably responses to climate warming. Ecol Appl. https://doi.org/10.2307/1941811
Hamada Y, Darung U, Limin SH, Hatano R (2013) Characteristics of the fire-generated gas emission observed during a large peatland fire in 2009 at Kalimantan, Indonesia. Atmos Environ 74:177–181
Hedberg P, Saetre P, Sundberg S, Rydin H, Kotowski W (2013) A functional trait approach to fen restoration analysis. Appl Veg Sci 16:658–666
Herbichowa M (2007) Eksperymentalna reintrodukcja gatunków z rodzaju Sphagnum. In: Herbichowa M, Pawlaczyk P, Stańko R (eds) Ochrona wysokich torfowisk batyckich na Pomorzu. Doświadczenia i rezultaty projektu LIFE 04/NAT/PL/00208 PLB BOGS. Wyd. Klub Przyrodników, Świebodzin, p 128–130
Higuera PE (2015) Taking time to consider the causes and consequences of large wildfires. Proc Natl Acad Sci USA 112:13137–13138
Hooijer A, Page S, Canadell JG, Silvius M, Kwadijk J, Wösten H, Jauhiainen J (2010) Current and future CO_2 emissions from drained peatlands in Southeast Asia. Biogeosciences 7(5):1505–1514
Ilnicki P, Iwaniszyniec P (2002) Emmisions of greenhouse gases (GHG) from peatland in Restoration of carbon sequestrating capacity and biodiversity in abandoned grassland on peatland in Poland, Wyd. Akademii Rolniczej w Poznaniu: 19–55
IPCC (2013) In: Stocker TF, Qin D, Plattner GK, Tignor M, Allen SK, Boschung J, Nauels A, Xia Y, Bex V, Midgley PM (2013) Climate Change 2013: The Physical Science Basis.

Contribution of Working 25 Group I to the Fifth Assessment Report of the Intergovernmental Panel on Climate Change. IPCC, AR5:1535

Ise T, Dunn AL, Wofsy SC, Moorcroft PR (2008) High sensitivity of peat decomposition to climate change through water-table feedback. Nat Geosci 1:763–766

Jassey VE, Signarbieux C, Hattenschwiler S, Bragazza L, Buttler A, Delarue F, Fournier B, Gilbert D, Laggoun-Defarge F, Lara E, Mills RT, Mitchell EA, Payne RJ, Robroek BJ (2015) An unexpected role for mixotrophs in the response of peatland carbon cycling to climate warming. Scientific reports 5:16931

Jassey VEJ, Lamentowicz M, Bragazza L, Hofsommer ML, Mills RTE, Buttler A, Signarbieux C, Robroek BJM (2016) Loss of testate amoeba functional diversity with increasing frost intensity across a continental gradient reduces microbial activity in peatlands. Europ J Protistol 55 (B):190–202

Joosten H, Tanneberger F, Moen A (2017) Mires and peatlands of Europe. Schweizerbart Science Publishers, Germany

Kajukalo K, Fialkiewicz-Koziel B, Galka M, Kolaczek P, Lamentowicz M (2016) Abrupt ecological changes in the last 800 years inferred from a mountainous bog using testate amoebae traits and multi-proxy data. Europ J Protistol 55:165–180

Keddy PA (2002) Wetland Ecology: Principles and Conservation. Cambridge University Press, UK

Kettridge N, Turetsky MR, Sherwood JH, Thompson DK, Miller CA, Benscoter BW, Flannigan MD, Wotton BM, Waddington JM (2015) Moderate drop in water table increases peatland vulnerability to post-fire regime shift. Scientific Reports 5:8063

Kleinen T, Brovkin V, Munhoven G (2016) Climate of the Past, Modelled interglacial carbon cycle dynamics during the Holocene, the Eemian and Marine Isotope Stage (MIS) 11. Clim Past 12:2145–2160

Klimkowska A, Dzierża P, Kotowski W, Brzezińska K (2010) Methods of limiting willow shrub re-growth after initial removal on fen meadows. J Nat Conserv 18:12–21

Kotowski W, Ackermann M, Grootjans AP, Klimkowska A, Rossling H, Wheeler B (2016) Restoration of temperate fens: matching strategies with site potential. In: Bonn A, Allott T, Evans M, Joosten H (eds) Peatland Restoration and Ecosystem Services. Science, p 172–193

Köchy M, Hiederer R, Freibaue A (2015) Global distribution of soil organic carbon—Part 1: Masses and frequency distributions of SOC stocks for the tropics, permafrost regions, wetlands, and the world. Soil. https://doi.org/10.5194/soil-1-351-2015

Kuhry P (1994) The role of fire in the development of sphagnum-dominated peatlands in western boreal Canada. J Ecol 82(4):899–910

Kulczyński S (1949) Peatbogs of Polesie Mémoires de l 'Académie Polonaise des Sciences et des Lettres. B Sci Nat 15:1–356

Lamentowicz M, Tobolski K, Mitchell EAD (2007) Palaeoecological evidence for anthropogenic acidification of a kettle-hole peatland in northern Poland. The Holocene 17(8):1185–1196

Lamentowicz M, Milecla K, Gałka M, Cedro A, Pawytla J, Piotrowska N, Lamentowicz Ł, van der Knaap (2008) Climate and human induced hydrological change since AD 800 in an ombrotrophic mire in Pomerania (N Poland) tracked by testate amoebae, macro-fossils, pollen and tree rings of pine. Boreas 38:214–229

Lamentowicz M, Mueller M, Gałka M, Barabach J, Milecka K, Goslar T, Binkowski M (2015) Reconstructing human impact on peatland development during the past 200 years in CE Europe through biotic proxies and X-ray tomography. Quatern Int 357:282–294

Lamentowicz M, Słowińska S, Słowiński M, Jassey VEJ, Chojnicki BH, Reczuga MK, Zielińska M, Marcisz K, Lamentowicz Ł, Barabach J, Samson M, Kołaczek P, Buttler A (2016) Combining short-term manipulative experiments with long-term palaeoecological investigations at high resolution to assess the response of Sphagnum peatlands to drought, fire and warming. Mires and Peat 18:1–17

Lappalainen E (1996) General review on world peatland and peat resources. In: Lappalainen E (ed) Global Peat Resources. International Peat Society and Geological Survey of Finland, Jyska, Finland, pp 53–56

Limpens J, Berendse F, Blodau C, Canadell JG, Freeman C, Holden J, Roulet N, Rydin H, Schaepman-Strub G (2008) Peatlands and the carbon cycle: from local processes to global implications—a synthesis. Biogeosciences 7:3517–3530

Loisel J, Yu Z, Beilman D, Philip C, Jukka A, David A, Andersson S, Fiałkiewicz-Kozieł B, Barber K, Belyea L, Bunbury J, Chambers F, Charman D, de Vleeschouwer F, Finkelstein S, Garneau M, Hendon D, Holmquist J, Hughes P, Jones M, Klein E, Kokfelt U, Korhola A, Kuhry P, Lamarre A, Lamentowicz M, Large D, Lavoie M, MacDonald G, Magnan G, Gałka M, Mathijssen P, Mauquoy D, McCarroll J, Moore T, Nichols J, O'Reilly B, Oksanen P, Peteet D, Rchard P, Robinson S, Rundgren M, Sannel B, Tuittila E-S, Turetsky M, Valiranta M, van der Linden M, van Geel B, van Bellen S, Vitt D, Zhao Y, Zhou W (2014) A database and synthesis of existing data for northern peatland soil properties and Holocene carbon accumulation. The Holocene 24:1028–1042

Maćkowiak M, Michalak A (2008) Biologia: Jedność i różnorodność. Wydawnictwo Szkolne PWN, Warszawa, pp 269–271

Main Report (2007) Assessment on Peatlands, Biodiversity and Climate change, Main Report. Global Environment Centre, Kuala Lumpur & Wetlands International, Wageningen, ISBN 978-983-43751-0-2

Marcisz K, Lamentowicz L, Slowinska S, Slowinski M, Muszak W, Lamentowicz M (2014) Seasonal changes in *Sphagnum* peatland testate amoeba communities along a hydrological gradient. Eur J Protistol 50:445–455

Marcisz K, Tinner W, Colombaroli D, Kołaczek P, Słowiński M, Fiałkiewicz-Kozieł B, Łokas E, Lamentowicz M (2015) Long-term hydrological dynamics and fire history during the last 2000 years in CE Europe reconstructed from a high-resolution peat archive. Quat Sci Rev 112:138–152

Matthews GVT (1993) The Ramsar Convention on wetlands: its history and development. Ramsar Convention Bureau, Gland, Switzerland

Mauquoy D, Yeloff D (2007) Raised peat bog development and possible responses to environmental changes during the mid- to late-Holocene. Can the palaeoecological record be used to predict the nature and response of raised peat bogs to future climate change? Biodivers Conserv. https://doi.org/10.1007/s10531-007-9222-2

Mäkilä M, Saarnisto M (2008) Carbon accumulation in boreal peatlands during the holocene—impacts of climate variations. In: Strack M (ed) Peatlands and Climate Change. International Peat Society, Finland

Miettinen J, Hooijer A, Vernimmen R, Liew SC, Page SE (2017) From carbon sink to carbon source: extensive peat oxidation in insular Southeast Asia since 1990. Environmental Research Letters 12

Milecka K, Kowalewski G, Fiałkiewicz-Kozieł B, Gałka M, Lamentowicz M, Chojnicki BH, Goslar T, Barabach J (2016) Hydrological changes in the Rzecin peatland (Puszcza Notecka, Poland) induced by anthropogenic factors: Implications for mire development and carbon sequestration. The Holocene. https://doi.org/10.1177/0959683616670468

Moore TR, Roulet NT, Waddington JM (1998) Uncertainty in predicting the effect of climate change on the carbon cycling of Canadian peatlands. Clim Change 40:229–245

Moore S, Evans CD, Page SE, Garnett MH, Jones TG, Freeman C, Hooijer A, Wiltshire AJ, Limin SH, Gauci V (2013) Deep instability of deforested tropical peatlands revealed by fluvial organic carbon fluxes. Nature. https://doi.org/10.1038/nature11818

Mulot M, Marcisz K, Grandgirard L, Lara E, Kosakyan A, Robroek BJ, Lamentowicz M, Payne RJ, Mitchell EA (2017) Genetic Determinism vs. Phenotypic Plasticity in Protist Morphology. The Journal of Eukaryotic Microbiology. https://doi.org/10.1111/jeu.12406

Natura (2000) http//:www.ec.europa.eu/environment/nature/natura2000/. Accessed on 1 Aug 2017

Nature protection (2017) https://pl.wikipedia.org/wiki/Ochrona_przyrody_w_Polsce. Accessed on 1 Aug 2017

Natural Resources Canada (2016) http://www.nrcan.gc.ca/forests/climate-change/forest-carbon/13103. Accessed on 1 Aug 2017

Page SE, Siegert F, Rieley JO, Boehm H-D, Jaya A, Limin S (2002) The amount of carbon released from peat and forest fires in Indonesia during 1997. Nature. https://doi.org/10.1038/nature01131

Page S, Hoscilo A, Langner A, Tansey K, Siegert F, Limin S, Rieley J (2009a) Tropical peatland fires in Southeast Asia. In: Cochrane MA (ed) Tropical fire ecology: climate change, land use, and ecosystem dynamics. Springer-Praxis Books, Heidelberg, pp 263–287

Page S, Hosciło A, Wösten H, Jauhiainen J, Silvius M, Rieley J, Ritzema H, Tansey K, Graham L, Vasander H, Limin S (2009b) Restoration ecology of lowland, tropical peatlands in southeast, asia: current knowledge and future, research directions. Ecosystems 12:888–905

Page SE, Rieley JO, Banks CJ (2011) Global and regional importance of the tropical peatland carbon pool. Glob Change Biol 17(2):798–818

Petrescu AMR, Lohila A, Tuovinen J-P, Baldocchi DD, Desai AR, Roulet NT, Vesala T, Dolman AJ, Oechel WC, Marcolla B, Friborg T, Rinne J, Matthes JH, Merbold L, Meijide A, Kiely G, Sottocornola M, Sachs T, Zona D, Varlagin A, Lai DYF, Veenendaal E, Parmentier F-JW, Skiba U, Lund M, Hensen A, van Huissteden J, Flanagan LB, Shurpali NJ, Grünwald T, Humphreys ER, Jackowicz-Korczyński M, Aurela MA, Laurila T, Grüning C, Chiara AR, Corradi CAR, Schrier-Uijl AP, Christensen TR, Tamstorf MP, Mastepanov M, Martikainen PJ, Verma SB, Bernhofer C, Cescatti A (2015) The uncertain climate footprint of wetlands under human pressure. Proc Natl Acad Sci. https://doi.org/10.1073/pnas.1416267112

Postel S (1997) Last oasis: facing water scarcity. WW Norton & Co, New York, p 239

Randerson JT, Liu H, Flanner MG, Chambers SD, Jin Y, Hess PG, Pfister G, Mack MC, Treseder KK, Welp LR, Chapin FS, Harden JW, Goulden ML, Lyons E, Neff JC, Schuur EAG, Zender CS (2006) The Impact of Boreal Forest Fire on Climate Warming. Science 314:1130–1132

Rieley JO, Ahmad-Shah A-A, Brady MA (1996) The extent and nature of tropical peat swamps. In: Maltby E, Immirzi CP, Safford RJ (eds) Tropical lowland peatlands of southeast asia. IUCN, Gland, Switzerland, pp 17–53

Rooney RC, Bayley SE, Schindler DW (2011) Oil sands mining and reclamation cause massive loss of peatland and stored carbon. PNAS 109(13):4933–4937

Ruddiman WF (2003) The anthropogenic greenhouse era began thousands of years ago. Clim Change 61(3):261–293

Rydin H, Jeglum JK (2013) The biology of peatlands. Oxford University Press, UK

Sillasoo Ü, Väliranta M, Tuittila E-S (2011) Fire history and vegetation recovery in two raised bogs at the Baltic Sea. J Veg Sci 22:1084–1093

Słowińska S, Słowiński M, Lamentowicz M (2010) Relationships between Local climate and hydrology in *Sphagnum* Mire: implications for Palaeohydrological studies and ecosystem management. Pol J Env Stud 19:779–787

Strack M (2008) Peatlands and climate change. International Peat Society, Finland

The Guardian (2016) https://www.theguardian.com/environment/2016/may/11/canada-wildfire-environmental-impacts-fort-mcmurray. Accessed on 5 Aug 2017

Tobolski K (2012) Ochrona europejskich torfowisk, Współczesne Problemy Kształtowania i Ochrony Środowiska. In: Łachacz A (ed) Monografie nr 3p, 2012 Wybrane problemy ochrony mokradeł, Olsztyn

Tropical peatlands (2017) University of Helsinki. http://blogs.helsinki.fi/jyjauhia/. Accessed on 5 Aug 2017

Tuittila ES, Vasander H, Laine J (2000) Impact of rewetting on the vegetation of a cut-away peatland. Vegetation science. https://doi.org/10.2307/1478999

Turetsky M, Wieder K, Halsey L, Vitt D (2002) Current disturbance and the diminishing peatland carbon sink. Geographical Research Letters. https://doi.org/10.1029/2001GL014000

Turetsky MR, Donahue WF, Benscoter BW (2011) Experimental drying intensifies burning and carbon losses in a northern peatland. Nat Commun 2:514

Turetsky MR, Benscoter B, Page S, Rein G, van der Werf GR, Watts A (2015) Global vulnerability of peatlands to fire and carbon loss. Nature Geosci 8:11–14

Turunen J, Tomppo E, Tolonen K, Reinikainen A (2002) Estimating carbon accumulation rates of undrained mires in Finland–application to boreal and subarctic regions. The Holocene 12(1): 69–80

van der Werf GR, Randerson JT, Giglio L, Collatz GJ, Mu M, Kasibhatla PS, Morton DC, DeFries RS, Jin Y, van Leeuwen TT (2010) Global fire emissions and the contribution of deforestation, savanna, forest, agricultural, and peat fires (1997–2009). Atmos Chem Phys. https://doi.org/10.5194/acp-10-11707-2010

Whiting GJ, Chanton JP (2001) Greenhouse carbon balance of wetlands: methane emission versus carbon sequestration. Tellus B: Chemical and Physical Meteorology 53(5):521–528

World Energy Council (2013) World energy resources: peat. https://www.worldenergy.org/wp-content/uploads/2013/10/WER_2013_6_Peat.pdf. Accessed on 5 Aug 2017

Wösten JHM, Ismail AB, van Wijk ALM (1997) Peat subsidence and its practical implications: a case study in Malaysia. Geoderma 78:25–36

Yale Environment (2017) http://e360.yale.edu/features/can-we-discover-worlds-remaining-peatlands-in-time-to-save-them. Accessed on 5 Aug 2017

Yokelson RJ, Susott R, Ward DE, Reardon J, Griffith DWT (1997) Emissions from smoldering combustion of biomass measured by open-path Fourier transform infrared spectroscopy. J Geophysical Res: Atmospheres 102(D15):18865–18877

Yu Z (2007) Holocene carbon accumulation of fen peatlands in Boreal Western Canada: a complex ecosystem response to climate variation and disturbance. Ecosystems. https://doi.org/10.1007/s10021-006-0174-2

Yu Z, Beilman DW, Jones MC (2009) Sensitivity of Northern Peatland carbon dynamics to holocene climate change. In: Baird AJ, Belyea LR, Comas X, Reeve AS, Slater LD (eds) Carbon cycling in Northern Peatlands. American Geophysical Union, Washington, D. C. https://doi.org/10.1029/2008GM000822

Yu Z, Beilman DW, Frolking S, MacDonald GM, Roulet NT, Camill P, Charman DJ (2011) Peatlands and their role in the global carbon cycle. Eos, Trans Am Geophys Union 92(12): 97–98

Yu Z, Campbell ID, Campbell C, Vitt DH, Bond GC, Apps MJ (2003) Carbon sequestration in western Canadian peat highly sensitive to Holocene wet-dry climate cycles at millennial timescales. The Holocene 13(6):801–808

Yu Z, Loisel J, Brosseau DP, Beilman DW, Hunt SJ (2010) Hydrology and land surface studies, global peatland dynamics since the last glacial maximum. Geophys Res Lett https://doi.org/10.1029/2010GL043584

Zech R, Huang Y, Zech M, Tarozo R, Zech W (2011) High carbon sequestration in Siberian permafrost loess-paleosols during glacials. Clim Past 7:501–509

Zerbe S, Steffenhagen P, Parakenings K, Timmermann T, Frick A, Gelbrecht J, Zak D (2013) Ecosystem service restoration after 10 Years of rewetting peatlands in NE Germany. Env Manag 51(6):1194–1209

Sources, Transport and Sinks of Radionuclides in Marine Environments

Kasper Zielinski

Abstract In this paper, major global sources of radionuclides are presented and discussed with respect to marine environments. It has been shown that the atmospheric fallout from nuclear weapons testing has been the most significant source of radionuclides in the oceans and seas. As a result, areas within the 0°–60° N latitude bands received the greatest loads of ^{137}Cs, which is related to the fact that the majority of nuclear test sites were located at these latitudes. The relative input from the Chernobyl disaster to the oceans was significantly lower than from the fallout from nuclear weapons tests. It has also been presented that in order to determine the potential impact of radionuclides on humans, with the continuous release of radionuclides to the oceans and the seas, the total influence from all sources is more significant than the initial concentration.

Keywords Radionuclides · Pollutant loadings · Chernobyl disaster
Marine environment · Baltic Sea · Cesium

1 Introduction

Radionuclides are naturally produced as a result of stellar nucleosynthesis and supernova explosions. Most of them decay quickly and are unobservable on Earth, others, such as uranium and thorium are found on Earth due to having a very long half-life.

Radionuclides can also be produced artificially by thermonuclear reactions during the explosion of nuclear weapons and by nuclear fission. Most of the anthropogenic radionuclides were introduced into the environment during the extensive nuclear weapons tests of the 1960s. With over 70% of the surface of the Earth being covered with water, it is obvious that the majority of radionuclides

K. Zielinski (✉)
University of Gdansk, Gdansk, Poland
e-mail: zielinski.kasper@gmail.com

T. Zielinski et al. (eds.), *Interdisciplinary Approaches for Sustainable Development Goals*, GeoPlanet: Earth and Planetary Sciences,
https://doi.org/10.1007/978-3-319-71788-3_13

released to the environment are found in the waters of oceans and seas and in their sediments (Zaborska et al. 2014).

Due to the variety of radionuclide sources and their further transport, dispersion and mixing in waters, the marine environment is loaded with radionuclides, which are specific for given regions.

There is significant interest in the study of the behavior of radionuclides, in ocean waters. The fate of radionuclides must be documented in order to learn whether they have an insignificant or adverse impact on the environment, including human health. Such knowledge provides the foundations for future assessments regarding potential releases. These releases may include unplanned scenarios, such as accidental releases of radionuclides from coastal nuclear facilities, nuclear waste disposal sites and nuclear fuel transport (IAEA 2000). Radionuclides also serve as important tracers of many environmental processes in the ocean and its sedimentary systems, such as the carbon cycle (Bond et al. 2013). Ocean sediments act as a major sink for radionuclides, however, they can also act as a source under specific circumstances (Zaborska et al. 2014).

Studies of isotope ratios provide information about the sources of radionuclides in water and sediments. Therefore, in order to estimate marine radioactivity in a particular area, information about the sources and oceanic processes must be available. The main source, i.e. global nuclear fallout, which is mainly related to nuclear weapons tests of the 1960s, is evaluated based on monitoring activities performed mainly on land (NATO 1998). Local levels of radionuclide contamination is related to discharges from nuclear fuel reprocessing plants or dumping of liquid and solid radioactive waste. However, soluble radionuclides are transported over vast distances with ocean currents (NATO 1998).

We now know that primary sources are significantly reduced today, but the result of the primary inputs of radionuclides may be manifested in a form of a diffuse secondary source (Hunt et al. 1990; Oughton and Day 1993). Beside soluble species, in the case of surface runoff from land ecosystems, two basic sources of transfer between sediments and waters may occasionally take place: transport of suspended sediments during floods and the transport of sediments with and in ice (Meese et al. 1995; Pfirman et al. 1995). Consequently, terrigenous sediments are introduced to sea waters, and may be further transformed and thus may affect the mobility of radionuclides.

The evaluation of radionuclide inputs from local sources and their distribution in various marine environments requires thorough studies about the present situation as well as knowledge about the past inputs of the main radionuclides of interest, i.e. ^{3}H, ^{14}C, ^{90}Sr, ^{137}Cs and Pu isotopes (IAEA 2005). Such knowledge also facilitates the creation of future models of radionuclide concentrations and their distribution in the world's oceans and seas.

In this paper, I describe the importance of radionuclide releases and their role in marine environments. I provide information about the main sources of radionuclides in the oceans and the seas. Finally, I describe the case of the Baltic Sea, the main sink of Cesium (^{137}Cs) released during the Chernobyl disaster.

2 Anthropogenic Sources of Marine Radioactive Contamination

The oceans and seas are sinks for radioactive particles originating from global fallout from the atmosphere, from liquid and solid waste dumps and indirectly from river runoff. Some radionuclides are of conservative character and remain soluble in the water, while other are insoluble or adhere to particles and thus, are moved, with time, to marine sediments (Borretzen and Salbu 2002).

Radionuclides are transferred between water and sediments in many ways. These processes can be physical (e.g. sedimentation), chemical (e.g. ion-exchange, polymerization, colloid aggregation) or biological (e.g. detritus) in nature. The physical transfer of contaminants from sediments to water occurs as a result of natural and/or anthropogenic resuspension of the sediments. This happens during e.g. floods, erosion or dredging. The chemical mobilization includes ion-exchange, leaching and dissolution. Finally, biological processes affect both the chemical and physical mobilization through e.g. bioturbation. The most significant global source of radionuclides in the oceans and seas is global fallout from atmospheric nuclear tests.

The Chernobyl disaster is among the most significant sources of anthropogenic radionuclides. In some areas, such as e.g. the Baltic Sea or the Black Sea, the concentration of ^{137}Cs in the water is closely related to the Chernobyl disaster (Ilus 2007; Tarasiuk et al. 1995). For these areas, concentration changes of ^{137}Cs are quite rapid. Additionally, water-borne discharges from nuclear reprocessing plants, mostly Sellafield in the United Kingdom and Cap de la Hague in France, have contributed above the world average (Charlesworth et al. 2006; Gulliver et. al. 2001), with Sellafield having a leading role in radionuclide contribution to the marine environment. Other sources, such as, sea dumping of nuclear waste, releases from nuclear power plants, satellite losses, sunken nuclear submarines, lost nuclear weapons and the use of radioisotopes in medicine, industry and science, altogether contribute significantly less (orders of magnitude) to the oceans and the seas than the three major radionuclide sources mentioned above.

Discharges of radiocesium from the Sellafield site had been considerably reduced in the late 1970s while the operation of the Thermal Oxide Reprocessing Plant (THORP) and the Enhanced Actinide Removal Plant (EARP), which were opened in 1994, have increased the release of some long-lived radionuclides, such as ^{137}Cs, ^{90}Sr or ^{99}Tc.

The Chernobyl disaster, which occurred in April 1986, resulted in an unprecedented release of radionuclides which were introduced to the atmosphere, dispersed and spread over great distances leading to the contamination of vast parts of Europe, mainly as a result of fallout, especially in the marine area west of Norway, in the Baltic Sea, and in the form of river run-off from contaminated land areas (Povinec et al. 2003).

The Arctic is another region in the Northern Hemisphere that was exposed to high amounts of radioactive contamination, mainly due to atmospheric nuclear tests (Cota et al. 2006). Additionally, between 1959 and 1991, the Soviet Union used the

Barents and Kara Seas as dump sites for nuclear waste, ranging from low level liquid waste to spent nuclear fuel. Radioactive discharges from former Soviet nuclear facilities are introduced into the drainage systems of the Ob River (Tomsk and Mayak) and the Yenisei River (Krasnoyarsk), both of which outflow to the Kara Sea (AMAP 2010).

No contribution from dumped radioactive waste has been observed in the open sea, and the estimations of the total releases from dumped nuclear waste show a rather small impact on the population (Fisher et al. 1999).

3 Radionuclides in the Global Ocean

3.1 Nuclear Weapons Testing

Atmospheric nuclear test explosions have affected the Earth on a global scale. The weapons tested between 1945 and 1951 had a yield ranging between 20 and 100 kilotons. Their testing resulted in atmospheric (mostly tropospheric) fallout. This means that the radioactive debris was captured in the troposphere, was not distributed globally, and fell-out or precipitated in the latitude bands relative to the testing areas (Livingstone and Povinec 2002). In the following years, (in 1952 and 1953, respectively) the USA and the Soviet Union developed and began testing thermonuclear weapons which could achieve much higher yields and produced radioactive fallout on a global scale. The intensity of these tests reached their peak between the years 1962 and 1964. Until the late 1960s the most significant amounts of radionuclides released due to nuclear weapons testing were related to the activities at test sites located around the Novaya Zemlya Island in the Arctic Ocean (80.89 Mt fission yield) and at the Bikini Atoll in the Pacific Ocean (42.2 Mt fission yield) (Aoyama and Hirose 2003). Other such facilities, though of a smaller scale, were located on the Enewetak and Johnston Atolls in the North Pacific (26 Mt), Christmas Island in the Indian Ocean (15.5 Mt), at Lop Nor salt lake in China (12.2 Mt), the Mururoa and Fangataufa Atolls in the South Pacific (6.1 Mt) and Semipalatinsk in Kazakhstan (3.7 Mt) (Aoyama and Hirose 2003).

The global radionuclide fallout pattern shows the atmospheric transport of debris from the stratosphere to the troposphere and the greatest transfer at mid-latitudes. It is obvious that the location of the particular test sites influenced the global precipitation pattern. It is clear that much lower amounts of radioactive fallout in the Southern hemisphere are related to the significantly lower numbers of test explosions that took place there, as well as to the limited northern and southern stratospheric exchange processes.

Table 1 shows the inputs of ^{137}Cs (using ^{90}Sr data and then multiplying by 1.6) from global fallout to all five oceans using the known values of fallout over land (Baumgartner and Reichel 1975; UN 2000). In order to obtain present day (year 2000s) information for the world oceans, the values in Table 1 should be decay-corrected by a factor of 0.40.

Table 1 ^{137}Cs input to the world ocean from global fallout in PBq

Ocean region	Pacific ocean	Atlantic ocean	Indian ocean	Arctic ocean	Total
N 90°–60°	2.1	16.3	0.0	7.4	25.8
N 60°–30°	114.6	91.7	0.0	0.0	206.2
N 30°–0°	105.3	49.3	21.3	0.0	175.8
S 0°–30°	42.6	16.5	23.4	0.0	82.4
S 30°–60°	41.3	24.6	36.0	0.0	101.9
S 60°–90°	4.8	2.7	3.4	0.0	10.9
Total	310.6	201.1	84.0	7.4	603.0

Source IAEA (2005)

It should be noted that the areas at the 0°–60° N latitude bands received the greatest loads of ^{137}Cs which, as mentioned above, is related to the majority of test sites being located at these latitudes. It is true for both the Atlantic and the Pacific Oceans—the largest oceans and thus the greatest sinks. In comparison, the Arctic Ocean is relatively free from radioactive contamination (AMAP 2010).

3.2 *Nuclear Reprocessing*

The aim of nuclear reprocessing is to recover uranium and plutonium from spent fuel for their further use in reactors. Three of the largest reprocessing plants are located in France (Cap de La Hague), the United Kingdom (Sellafield), and Japan (Tokai). On a global scale, Sellafield has been the largest source of such radioactive discharges to the marine environment. These discharges, mostly of ^{137}Cs, have been observed and measured in vast areas of the North Atlantic as well as in the Arctic Ocean (HELCOM 2010). The releases of ^{137}Cs from Cap de la Hague have reached up to 3% of the corresponding releases from the Sellafield site. The releases from the major European reprocessing facilities included 40 PBq ^{137}Cs and 6.5 PBq ^{90}Sr, 126 PBq ^{3}H, 1 PBq ^{99}Tc and 0.015 PBq ^{129}I (IAEA 2005). These releases have strongly contributed to the contamination of the world oceans, especially in the North Atlantic (30°–90°N). The discharge of ^{3}H from reprocessing is insignificant in comparison to that from global fallout. However, both sites, Sellafield and Cap de la Hague are key contributors of ^{99}Tc and ^{129}I to the oceans and the seas. These radionuclides, have very long half-lives (213,000 and 1,570,000 years, respectively) and thus, will remain in the marine environment for much longer than e.g. cesium.

3.3 *Nuclear Disasters*

Most nuclear disasters did not affect the global ocean directly, as these incidents usually lead to atmospheric pollution, therefore, the only input of radioactive particles to the seas and oceans was due to atmospheric deposition or river runoff. However, there have been a number of nuclear accidents which resulted in the significant input of radionuclides directly into marine environments.

Over the course of years there have been many nuclear incidents that occurred all around the world. Some of the most notable of these disasters are the two large-scale nuclear accidents in Kyshtym and Windscale, which took place in 1957, however they had little to no effect on the marine environment.

In 1964 the Transit 4A/B satellite equipped with a SNAP-9A nuclear generator failed to reach orbit and disintegrated in the atmosphere, releasing approximately 1 kg of ^{238}Pu, some of which entered the marine environment. As a result, water in the Southern Hemisphere showed an elevated ^{238}Pu/^{239,240}Pu ratio, in comparison to the ocean waters in the Northern Hemisphere (NAS 1971).

Four years later, in 1968, a B-52 bomber crashed into sea-ice 11 km west of the Thule Airbase in North-West Greenland. As a result, the local marine environment received about 1 TBq ^{239,240}Pu (Aarkorg et al. 1984).

Another source of disaster-related radioactive pollution involves nuclear submarine losses, such as e.g. the Russian “Kursk”, which sank in the Barents Sea in August 2000. However, the “Kursk” was recovered in 2001 and contamination had been prevented. The sinking of the Soviet “Komsomolets” was a different case. This nuclear submarine sank at a depth of 1700 m near Bear Island in the Norwegian Sea. The ship wreck is believed to produce an estimated 2.8 PBq ^{90}Sr and 3 PBq ^{137}Cs, while the warheads may contain an additional 16 TBq ^{239,240}Pu (JRNEG 1994). Detectably enhanced levels of ^{137}Cs have been observed in sea water in the vicinity of the wreck.

The most recent nuclear disaster occurred in Japan in March 2011. A huge tsunami destroyed three nuclear reactors in the Fukushima power plant and caused serious discharges of radionuclides to the atmosphere and ocean water. The atmospheric releases were transported with winds across the Northern Hemisphere, reaching Europe. Even though, the levels were very low with insignificant impact on human health, they were still high enough to have been detected between March and mid-May 2011 (HELCOM 2013). However, the radioactive fallout to the Baltic Sea resulting from the disaster was insignificant and rather undetectable in Baltic water and fish.

The most significant incident, resulting in the severe contamination of the Baltic Sea and the surrounding region was the 1986 Chernobyl disaster which is discussed in detail in Sect. 4.

3.4 Sea Dumping

Sea dumping was a common practice between the late 1940s and early 1960s, mainly in the Atlantic and Pacific Oceans. In 1967, about 0.3 PBq of solid waste was deposited at 5 km depth in the eastern Atlantic Ocean, as a result of international effort, led by the European Nuclear Energy Agency. Such international activities were carried out until 1982, when about 0.7 PBq α activity, 42 PBq β activity and 15 PBq tritium had been dumped in the North Atlantic (EU 1989; IAEA 1999). Insignificant releases of plutonium have been recorded in the dump areas (NEA 1996).

During the Soviet Union era, radioactive waste had been disposed in the Barents and Kara Seas. Estimates show that the total amount of waste dumped in the Arctic Seas amounted to approx. 90 PBq (Yablokov et al. 1993), and the monitoring activities proved (JNREG 1994) that the releases from the dump sites were insignificantly low. Further, model calculations have revealed that the total collective dose from these sites over the next 1000 years will be approx. 300 times less than the corresponding dose from the radioactive waste dump sites in the North Atlantic (NEA 1996).

3.5 *River Runoff*

Radionuclides which are deposited over land are transported to groundwater, lakes and rivers, and ultimately to the sea. Runoff is thus a secondary transfer of atmospheric radioactivity to the sea. About 9% of the ^{90}Sr inventory on land will be removed by runoff (Aarkrog 1979), while in the case of ^{137}Cs it is about 2% (Yamagata et al. 1963).

Global fallout is responsible for the release of 622 PBq ^{90}Sr and 948 PBq ^{137}Cs, and as 377 PBq ^{90}Sr and 603 PBq ^{137}Cs were deposited in the sea, this leaves the load deposited over land at 245 PBq ^{90}Sr and 345 PBq ^{137}Cs. The total deposition from the Chernobyl disaster has been estimated at ~10 PBq ^{90}Sr and ~85 PBq ^{137}Cs (UN 2000), of which, 10 PBq ^{90}Sr and 69 PBq ^{137}Cs were deposited over land.

This means that the biggest contributor to the radioactive contamination of the marine environment is global fallout. However, disasters and nuclear reprocessing plants/dump sites have much more impact on a local scale.

4 Chernobyl and ^{137}Cesium—Contamination in the Baltic Sea

There are three main sources of ^{137}Cs in the Baltic Sea, which are presented in Fig. 1.

The majority of the 100 PBq ^{137}Cs released during the Chernobyl disaster in 1986 was deposited over land. The total Chernobyl ^{137}Cs input to the world oceans has been estimated at about 16 PBq, with 10 PBq in the 30°–60° N latitude belt and 6 PBq in the 60°–90° N latitude belt (HELCOM 2013). Nearly all of the ^{137}Cs was

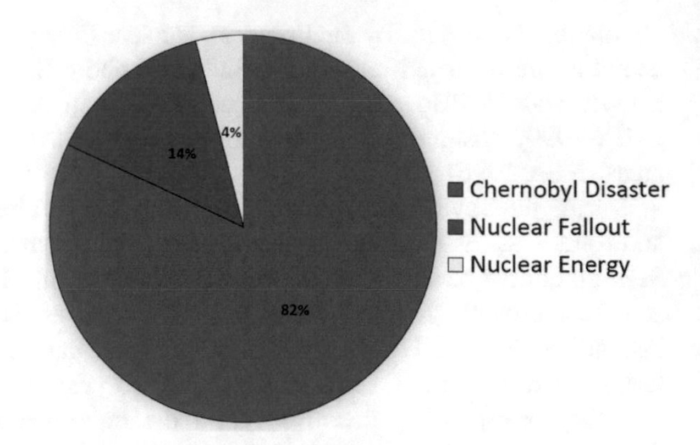

Fig. 1 Main sources of ^{137}Cs in the Baltic Sea

deposited in the North Atlantic and the Arctic Ocean. The relative Chernobyl disaster input to the oceans was thus significantly lower than that from nuclear weapons fallout. This is due to the fact that the Chernobyl disaster affected only the tropospheric layer of the atmosphere which resulted in the majority of radioactive material being deposited in the nearby European regions.

Relatively high amounts were deposited in the European marginal seas, including the Baltic Sea, which became the most contaminated, receiving 4.5 PBq ^{137}Cs (HELCOM 1995), in comparison to the 2–3 PBq ^{137}Cs in the Black Sea (EU 1994). The Black Sea outflow is the main source of "new" ^{137}Cs in the Mediterranean Sea (Evangeliou et al. 2009). The Mediterranean Sea was also directly affected by the Chernobyl fallout (EU 1989). The North Sea was estimated to have received 1.2 PBq ^{137}Cs from the Chernobyl disaster, while the North Atlantic received about 6 PBq ^{137}Cs. In 1987, surface sea water samples collected in the Greenland, Norwegian and Barents Seas and off the western coast of Norway and the Faroe Islands, were found to contain Chernobyl originated ^{137}Cs.

The estimated equivalent doses received by humans in the Baltic Sea region from Chernobyl fallout are significantly higher than those from the Fukushima event. Table 2 presents the results of the dose calculations. General distribution of ^{137}Cs in Baltic Sea sediments is presented in Fig. 2.

Table 2 Summary of the effects of Chernobyl and Fukushima fallouts on the Baltic Sea

Item	Quantity
Deposition od radiocesium over the Baltic Sea	4 Bq/m^2
Input of radiocesium to the Baltic Sea	1.6 TBq
Annual dose rate in 2011 due to Chernobyl fallout	0.9 μSv/a
Additional dose rate in 2011 due to Fukushima fallout	0.003 μSv/a

Source HELCOM (2013)

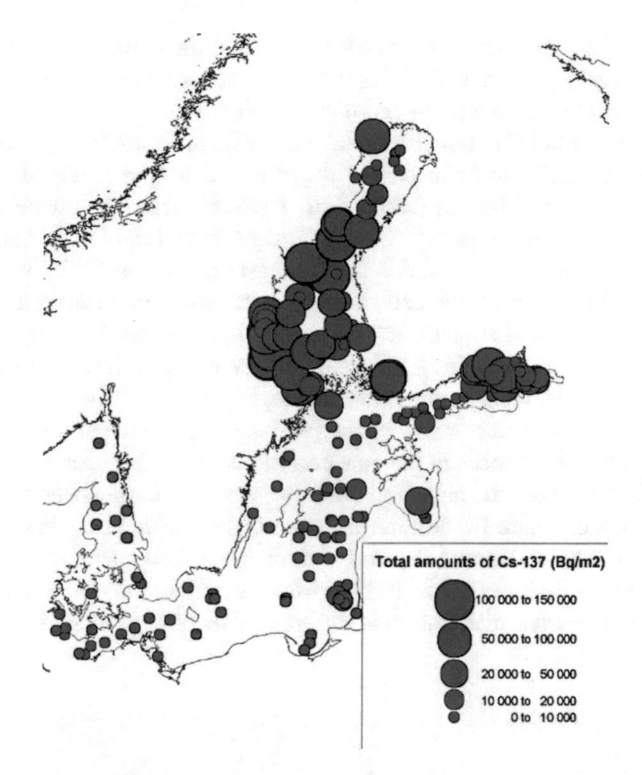

Fig. 2 Total amounts of ^{137}Cs (Bq m−2) in sediments at different sampling stations in the Baltic Sea at the beginning of the 2000s. Modified from HELCOM (2009)

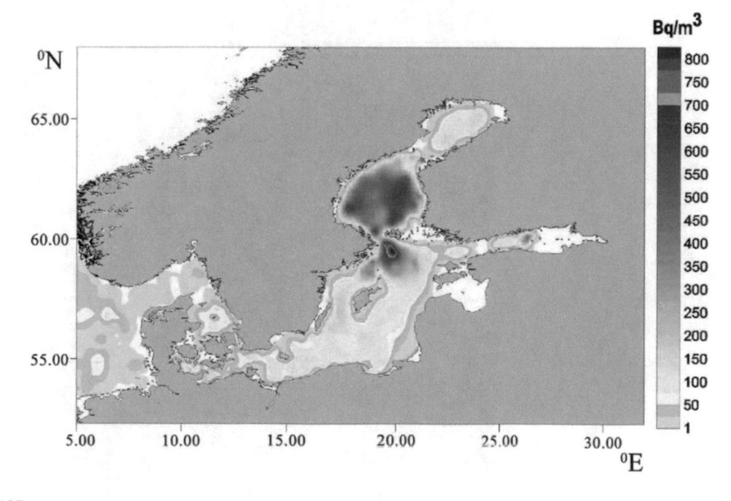

Fig. 3 ^{137}Cs contours of Baltic Sea surface waters after the Chernobyl accident (for the period of 1986–1988) as extracted from the GLOMARD database. Modified from IAEA (2005)

It is clear that sediments in the northern regions, including the Gulf of Bothnia, contain the largest quantities of ^{137}Cs. This is in accordance with the direct fallout distribution for the Baltic Sea, as presented in Fig. 3.

The accident at the Chernobyl Nuclear Power Plant deposited large quantities of radionuclides over the northern part of the Baltic Sea. The levels of ^{137}Cs in the Baltic Sea were defined by this fallout from 1986 onwards (Davuliene et al. 2007). Prior to the Chernobyl disaster, ^{137}Cs levels were related only to global fallout and the inflows of North Sea waters. After 1986, vast amounts of ^{137}Cs were deposited in the Baltic fine grained sediments, however, the amounts measured in the water have decreased by about a factor of 2 since the disaster (Outola 2010). The activity ratio of ^{137}Cs/^{90}Sr for the Baltic Sea is unusual due to inputs from a large number of rivers. However, this mechanism is almost negligible for ^{137}Cs (HELCOM 2010).

The Baltic Sea is characterized by the permanent presence of a halocline which causes significant differences in the concentrations of ^{137}Cs in surface water and in the deeper layers below the halocline. This is especially pronounced in the western part of the Baltic, where the inflow of saline waters with lower levels of cesium occur below the halocline and the outflows of less saline surface waters transport larger amounts of cesium (IAEA 2005). However, mixing of water masses has led to a more homogeneous distribution in different parts of the Baltic over years. ^{137}Cs

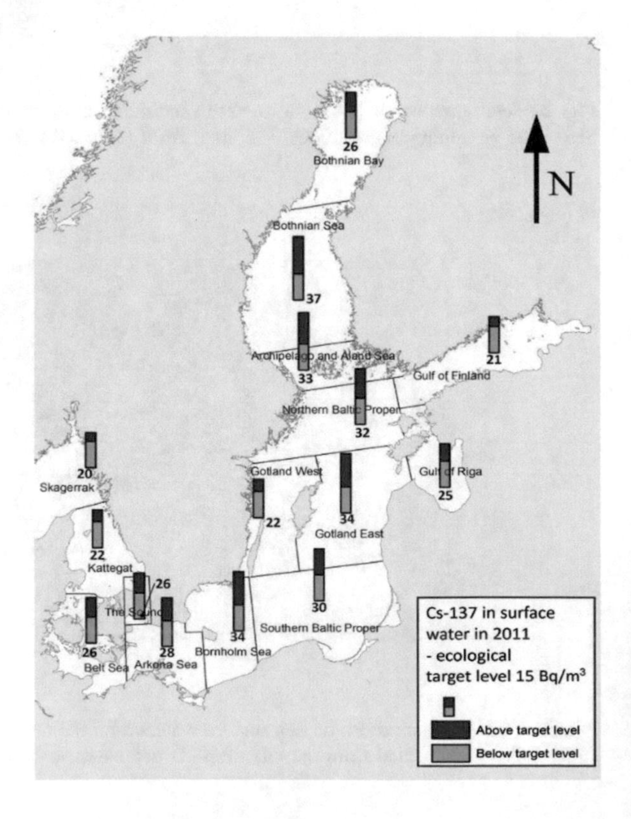

Fig. 4 ^{137}Cs concentrations (in Bq/m3) in surface water (sampling depth <=10 m) in 2011, as annual mean values by basin. Modified from HELCOM (2013)

concentrations (in Bq/m^3) in surface water (sampling depth <=10 m) in 2011 are presented as annual mean values in Fig. 4. Ecological target values have been calculated as average of pre-Chernobyl (1984–1985) cesium concentrations.

It is evident that all regions of the Baltic show elevated levels of cesium concentrations in surface waters.

Another, key problem with secondary cesium pollution of the Baltic Sea comes from the presence of this radionuclide in the smoke from wild forest fires in Belarus and Ukraine. A study of summer wind patterns for the Baltic Proper conducted by Zdun et al. (2011) indicate that around 50% of prevailing winds are of easterly and southerly directions. Therefore, they pass over the regions of Belarus and Ukraine. Zielinski et al. (2004) also reported that the majority of winds in summer come to the southern part of the Baltic Sea, from NE and SE directions, which confirms Zdun's observations.

There have been several large-scale forest fires in Belarus and Ukraine in the past two decades, most notably in the summer of 2002, spring/summer of 2008, and in 2010 and 2015—during which the fires were located within the Chernobyl exclusion zone. The large scale and long duration of the fires resulted in significant biomass burning, leading to the release of substantial quantities of radionuclides including ^{137}Cs into the atmosphere. However, data availability on the 2015 fires is still limited.

In March 2008, ^{137}Cs from forest fires covered the area of Belarus and a part of Russia, while another event in August distributed ^{137}Cs on a more local scale (HELCOM 2013). However, the 2002 and 2010 summer fires affected several European countries, including the Balkans, Turkey, and part of Italy (Fig. 5).

The highest exposure to the radioactive cloud from the 2002, 2008, and 2010 fires occurred over central Europe, with the most exposure of the Baltic occurring in 2002. The deposition of ^{137}Cs was largest in 2002 compared to 2008 and 2010,

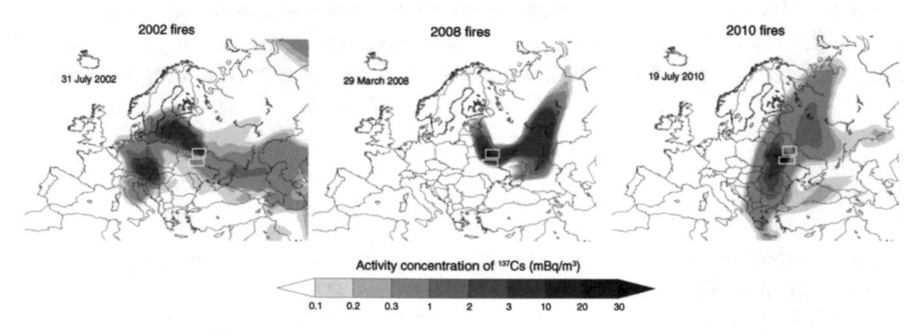

Fig. 5 Examples of ^{137}Cs emissions in the contaminated forests of Belarus and Ukraine after fire events in 2002, 2008, and 2010. The white boxes denote the location of the radioactive forests of Chernobyl (Ukraine) and Belarus. Modified from Evangeliou et al. (2015)

however, the concentrations were relatively low with insignificant impact on human health (HELCOM 2013).

5 Summary and Conclusions

Over five decades of research provides firm evidence that in terms of anthropogenic contribution of radionuclides to the oceans and seas, atmospheric fallout from nuclear weapons testing has been, and continues to be the most significant source. It has also been proven that in determining the potential impact of both natural and anthropogenic radionuclides on humans, with the continuous release of radionuclides to the oceans and the seas, the total influence from all sources is more significant than the initial concentration.

The key parameters used in the assessment of radionuclide contamination in marine environments are as follows:

- The cumulative effect of radionuclides has the greatest impact on the environment, therefore the proper classification of all sources of radioactivity is required.
- Bottom currents which influence the short-term dispersion of radionuclides.
- Interactions between water and sediments. The distribution pattern of radionuclides in water influences their further transfer.
- The behavior and properties of radionuclides in the marine food chain. Due to the possibility of bio-accumulation.

In regard of the Baltic Sea, the sources and parameters are different than in the case of the global ocean. There is a constant exchange of ^{137}Cs between the Baltic Sea and the North Sea through the Danish Straits. Radioactive discharges from nuclear power plants into the Baltic Sea area are relatively small and only detectable in the vicinity of the source. The main sources of radionuclides in the Baltic Sea are global fallout and local events, e.g. Chernobyl, or more recently, releases of radionuclides as a result of wild fire outbreaks in the areas surrounding Chernobyl.

References

Aarkorg A (1979) Environmental studies on radioecological sensitivity and variability with special emphasis on the fallout nuclides ^{90}Sr and ^{137}Cs. Risø-R-437, 1&2

Aarkorg A, Dahlgaard H, Nilsson K, Holm E (1984) Studies of plutonium and americium at Thule. Greenland Health Phys 46(1984):29–44

AMAP (2010) Radioactivity in the Arctic. Arctic Monitoring and Assessment Programme (AMAP), Oslo

Aoyama A, Hirose K (2003) Temporal variation of Cs-137 water column inventory in the North Pacific since the 1960s. J Environ Radioact 69(1–2):107–117

Baumgartner A, Reichel E (1975) The world water balance. Elsevier Science Publishing Company, Amsterdam and Oxford

Bond TC, Doherty S, Fahey D, Forster PM, Berntsen T, DeAngelo BJ, Flanner MG, Ghan S, Kärcher B, Koch D, Kinne S, Kondo Y, Quinn PK, Sarofim MC, Schultz MG, Schulz M, Venkataraman C, Zhang H, Zhang S, Bellouin N, Guttikunda SK, Hopke PK, Jacobson MZ, Kaiser JW, Klimont Z, Lohmann U, Schwarz JP, Shindell D, Storelvmo T, Warren SG, Zender CS (2013) Bounding the role of black carbon in the climate system: a scientific assessment. J Geophys Res Atmos 118:5380–5552

Borretzen P, Salbu B (2002) Fixation of Cs to marine sediments estimated by a stochastic modelling approach. J Environ Radioact 61(1):1–20

Charlesworth ME, Service M, Gibson CE (2006) The distribution and transport of Sellafield derived Cs-137 and Am-241 to western Irish Sea sediments. Sci Total Environ 354(1):83–92

Cota GF, Cooper LW, Darby DA, Larsen IL (2006) Unexpectedly high radioactivity burdens in ice-rafted sediments from the Canadian Arctic Archipelago. Sci Total Environ 366(1):253–261

Davuliene L, Tarasiuk N, Spirkauskaite N, Trinkunas G, Valkunas L (2007) ^{137}Cs activity distribution in the Lithuanian coastal waters of the Baltic Sea. Oceanologia 49(1):71–90

European Commission (1989) The radiological exposure of the population of the European community from radioactivity in north European marine waters. Project MARINA, Report EUR 12483, Luxembourg

European Commission (1994) The radiological exposure of the population of the European community from radioactivity in the Mediterranean Sea, Project MARINAMED, Report EUR 15564, Luxembourg

Evangeliou N, Florou H, Bokoros P, Scoullos M (2009) Temporal and spatial distribution of Cs-137 in Eastern Mediterranean Sea. Horizontal and vertical dispersion in two regions. J Environ Radioact 100(8):626–636

Evangeliou N, Balkanski Y, Cozic A, Hao WM, Mouillot F, Thonicke K, Paugam R, Zibtsev S, Mousseau TA, Wang R, Poulter B, Petkov A, Yue C, Cadule P, Koffi B, Kaiser JW, Møller AP (2015) Fire evolution in the radioactive forests of Ukraine and Belarus: future risks for the population and the environment. Ecol Monogr 85(1):49–72

Fisher NS, Fowler SW, Boisson F, Carroll J, Rissanen K, Salbu B, Sazykina TG, Sjoeblom KL (1999) Radionuclide bioconcentration factors and sediment partition coefficients in Arctic Seas subject to contamination from dumped nuclear wastes. Environ Sci Technol 33(12):1979–1982

Gulliver P, Cook GT, MacKenzie AB, Naysmith P, Anderson R (2001) Transport of Sellafield-derived C-14 from the Irish Sea through the North Channel. Radiocarbon 43 (2B):869–877

HELCOM (2009) Radioactivity in the Baltic Sea, 1999–2006. Helsinki Commission, Baltic Marine Environment Protection Commission

HELCOM (2010) Hazardous substances in the Baltic Sea—an integrated thematic assessment of hazardous substances in the Baltic Sea. Balt Sea Environment Proceedings No. 120B

HELCOM (2013) HELCOM Core Indicator of hazardous substances radioactive substances. Radioactive substances—Caesium-137 in fish and surface waters

Helsinki Commission, Baltic Marine Environment Protection Commission, HELCOM (1995) Radioactivity in the Baltic Sea 1984–1991, Balt Sea Environ Proc No. 61. ISSN 0357-2994

Hunt GJ, Kershaw PJ (1990) Remobilisation of artificial radionuclides from the sediment of the Irish sea. J Radiol Prot 10:147–152

IAEA (2000) Global Marine Radioactivity Database (GLOMARD). International Atomic Energy Agency: IAEA TECDOC Series No. 1146, Vienna

IAEA (2005) Worldwide marine radioactivity studies (WOMARS) Radionuclide levels in oceans and seas. IAEA-TECDOC-1429

Ilus E (2007) The Chernobyl accident and the Baltic Sea. Boreal Environ Res 12:1–10

International Atomic Energy Agency (IAEA) (1999) Inventory of radioactive waste disposals at sea, IAEA-TECDOC-1105, vol 127. IAEA, Vienna

Joint Russian-Norwegian Expert Group, JRNEG (1994) Radioactive contamination at dumping sites for nuclear wastes in the Kara Sea, results from the 1993 expedition. NRPA, Østerås

Livingston HD, Povinec PP (2002) A millennium perspective on the contribution of bomb fallout radionuclides to ocean science. Health Phys 82(5):656–668

Meese D, Cooper L, Larsen IL, Tucker W, Reimnitz W, Grebmeier J (1995) Cesium-137 contamination in Arctic Sea ice. In: Strand P, Cooke P (eds) Environmental radioactivity in the Arctic, pp 195–198. NRPA, Østerås

National Academy of Sciences, NAS (1971) Radioactivity in the marine environment, National Academy of Sciences, Washington, D.C. ISBN 0-309-01865-X

NATO—North Atlantic Treaty Organization (1998) Cross-border environmental problems emanating from defence-related installations and activities. Phase II: 1995–1998, final report vol 2. Radioactive contamination of rivers and transport through rivers, deltas and estuaries to the sea, Rep. No. 225, 105 pp

Nuclear Energy Agency (NEA) (1996) Co-ordinated research and environmental surveillance programme related to sea disposal of radioactive waste. CRESP final report 1981–1995, OECD, Paris

Oughton DH, Day JP (1993) Determination of cesium, rubidium and scandium in biological and environmental materials by neutron activation analysis. J Radioanal Nucl Chem 174 (1993):177–185

Outola I (2010) Total amounts of the artificial radionuclide cesium-137 in Baltic Sea sediments. HELCOM Indicator Fact Sheets 2010

Pfirman SL, Kogeler JW, Anseme B (1995) Transport of radionuclides from the Kara Sea: potential "shortcuts" in space and time. In: Strand P, Cooke A (eds) Environmental radioactivity in the Arctic, pp 191–194. NRPA, Østerås

Povinec PP, Bailly du Bois P, Kershaw PJ, Nies H (2003) Temporal and spatial trends in the distribution of ^{137}Cs in surface waters of Northern European Seas—a record of 40 years of investigations. Deep-Sea Res II 50:2785

Tarasiuk N, Spirkauskaite N, Stelingis K, Lujaniene G, Lujanas V (1995) Evaluation of the ^{137}Cs residence time in the Curonian Gulf after the Chernobyl fallout. Atmos. Phys. 2:22–33

United Nations Scientific Committee on the Effects of Ionizing Radiation (2000) Sources and effects of ionizing radiation. UN, New York

Yablokov AV et al (1993) Facts and problems related to radioactive waste disposal in sea adjacent to the territory of the Russian Federation. Small World Publishers, Moscow

Yamagata N, Matsuda S, Kodaira K (1963) Run-off of cesium-137 and strontium-90 from rivers. Nature 200(1963):668–669

Zaborska A, Winogradow A, Pempkowiak J (2014) Caesium-137 distribution, inventories and accumulation history in the Baltic Sea sediments. J Environ Radioact 127:11–25

Zdun A, Rozwadowska A, Kratzer S (2011) Seasonal variability in the optical properties of Baltic aerosols. Oceanologia 53(1):7–34

Zielinski T, Petelski T, Strzalkowska A, Pakszys P, Makuch P (2016) Impact of wild forest fires in Eastern Europe on aerosol composition and particle optical properties. Oceanologia 58(1):13–24

Sea Spray Aerosol Fluxes in the Sea-Air Boundary Layer—Description of Currently Used Methods and Review of Recent Achievements

Katarzyna Dziembor

Abstract Marine aerosols are small particles, solid and liquid, emitted to the atmosphere by a variety of factors related to the weather and the sea state. By its physical and chemical properties, the impact on the climate is significant, both on a global and local scale. The main purpose of this chapter is to introduce the reader to the concept of sea spray aerosol (SSA), its climate importance and way of emissions from the sea surface. Based on the recent comparision of aerosol emission from all the sources to the atmosphere, an advantage over those from other sources will be shown. After description of the idea of aerosol flux the currently used measurement methods will be presented, including references to the articles of the authors conducting such research, recent achievements on this area and our methods and results.

Keywords Sea spray Aerosol · Marine boundary Layer · SSA fluxes
Fluxes · Measurements methods

1 Introduction

Oceans cover circa 71% surface of Earth. About 92.2% of all water resources is contained in oceans and seas. The residue is located in glaciers, ice sheets and sea ice (2.15%), ground water (0.62%), rivers, streams and lakes (0.03%) (Stahler and Stahler 1992). The average depth of ocean is about 4 km. But in the case of SSA the most important is surface water. Considering the area covered by oceans, marine aerosols play a main role in emission to the atmosphere from all the sources, like volcano eruptions, terygenic, anthropogenic and biological factors. Adequate statements are presented in tables from the (Vignati et al. 2010; Boucher et al. 2013) articles—Tables 1 and 2, respectively. By definition aerosol is every single particle,

K. Dziembor (✉)
Faculty of Oceanography and Geography, University of Gdansk, Gdansk, Poland
e-mail: katarzyna.dziembor95@gmail.com

K. Dziembor
Institute of Oceanology Polish Academy of Sciences, Sopot, Poland

T. Zielinski et al. (eds.), *Interdisciplinary Approaches for Sustainable Development Goals*, GeoPlanet: Earth and Planetary Sciences,
https://doi.org/10.1007/978-3-319-71788-3_14

Table 1 Global natural aerosol emission (Tg/year, sulphur emissions in TgS/year), (Vignati et al. 2010)

Species	Source	References	Emission
POM	Fossil + bio fuels Biomass burning	Bond et al. (2004) van der Werf et al. (2004)	12.3 34.7
Black carbon	Fossil + bio fuels Biomass burning	Bond et al. (2004) van der Werf et al. (2004)	4.67 3.04
Sea salt	Wind driven	Gong (2003), Vignati et al. (2010)	6297
Dust	Wind driven	Dentener et al. (2006)	1776
SO_2	Industry, traffic, domestic, biomass burning, volcanos	Dentener et al. (2006), Cofala et al. (2007)	68.75
DMS	Marine	Kettle et al. (1999)	18.46[a]

POM particulate organic matter
[a]Year 2002–2003

solid or liquid, which is staying in the atmosphere. This definition does not include clouds. The emission of sea-surface droplets is one of the less known issues connected with sea-air interactions research. Sea salt, in the form of drops (so called marygenic aerosol), which gets to the higher part of the atmosphere, has great importance for correct description of radiation processes, boundary-layer cleaning processes of anthropogenic pollutions or cloud physics. The proper parameterization the aerosol flux from the sea surface is difficult issue. It is estimated, that sea salt emission varies between 1400 to 6800 Tg/year (Boucher et al. 2013), with relative uncertainty 80% (Tsigaridis et al. 2013). Such significant uncertainty in emission value estimation motivates to come up with more accurate methods of analysing sea-surface phenomena.

Table 2 Global natural emissions of aerosols and aerosol precursors in Tg/year (Boucher et al. 2013)

Source	Natural global	
	Min	Max
Sea spray	1400	6800
Inludung marine POA	2	20
Mineral dust	1000	4000
Terrestial PBAPs	50	1000
Including spores		28
Dimethylsulphide (DMS)	10	40
Monoterpenes	30	120
Isoprene	410	600
SOA production from all BVOCs	20	380

POA primary organic aerosol
PBAP primary biological aerosol particles
SOA secondary organic aerosol
BVOC biogenic volatile organic compounds

Referring to the tables, this is relatively easy to perceive the predominance of SSA over the aerosols emitted from other sources, irrelevant natural or anthropogenic. Oceans productivity is estimated at 50% of all natural sources of aerosol (Garbalewski 2000).

The importance of marine aerosols comes from proportion of emission from other sources. Salt is known for its hygroscopic properties, so the particles could became the cloud condensation nuclei (CCN), affect electromagnetic radiation throw light scattering, affect albedo, taking part in chemical reactions in the atmosphere, determine moisture and participate in global geochemical cycle. Moreover, marine aerosols play an important role in meteorology, oceanography, coastal ecology and has an impact on human health (Lewis and Schwartz 2004). That impact is determined by the particles diameter, their density in the air and time of deposition from the atmosphere.

2 Theory of SSA Flux and Recent Knowledge

SSA production is associated with the flux from the sea-surface, given in $m^{-2}\ s^{-1}$. According to literature there are several factors that cause emission. Aerosol production could be initiated, inter alia, by the wind over the ocean surface, leading to waves formation. Whitecaps, which appear on the waves, are exposed to horizontal wind action. Small particles are torn and float to the atmosphere as spume drops (Fig. 1).

Breaking wave is injecting the gas molecules under the water surface. Air particles are moving upwardly to the surface, where they form bubbles, existing due to the surface tension forces. Subsequently, when the bubbles burst, fragments of thin film get into the atmosphere as small droplets of diameter from 1 to 20 μm. The empty space is closing, and during this process jet drops of diameter ~ 100 μm, from the bottom of former bubble, appear and get into the atmosphere (Garbalewski 2000). The entire process is clearly presented on Fig. 2.

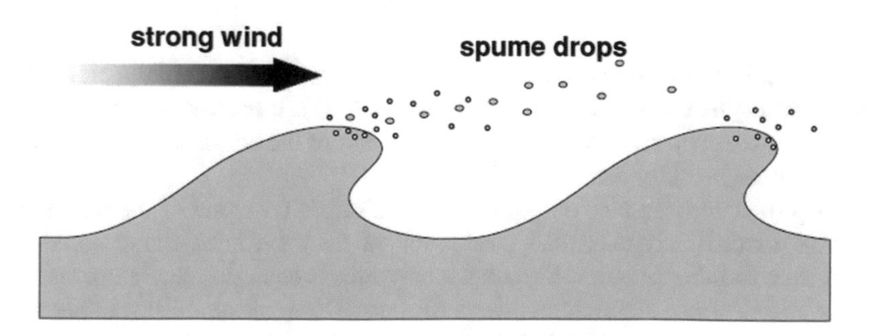

Fig. 1 Scheme of spume drops generation (Massel 2007)

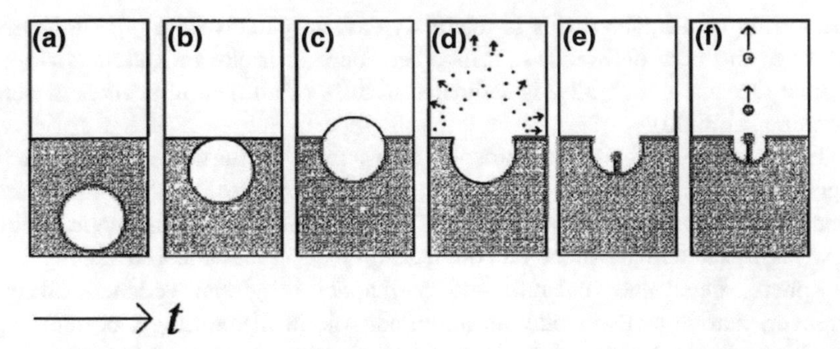

Fig. 2 Drop formation from bubble bursting. **a–c** Rise of the bubble to the surface and formation of the bubble film or cap. **d** Production of film drops from the bursting of the bubble film. **e** Shrinkage of the cavity left by the bubble and formation of a jet rising from the center of this cavity. **f** Further decrease in size of the cavity as it is filled by the surrounding liquid, and formation of several jet drops from breakup of the jet (Lewis and Schwartz 2004)

The most intensive emission is recorded during the storm, when wind speed reaches a value higher than 10 m s^{-1}. Particles are moving upwardly to the higher part of the atmosphere, susceptible to the turbulent diffusion. Part of them are precipitated by the cause of gravity force or atmospheric fallout. The disparity between emission and deposition is given as:

$$\frac{D}{Dt}\bar{n}(r) = \frac{\partial}{\partial x_i}\left(\overline{n'(r)u_l'} + \overline{n(r)}V_d(r)\right) \tag{1}$$

where N is a number concentration of particles, r is radius of droplet and $\overline{n(r)}$ is a mean value of concentration (Petelski 2003).

Atmospheric stability is another factor, that can be determined by the vertical profiles of temperature, wind speed, relative humidity near the sea surface and wind friction velocity. It also can be parametrized by the Monin-Obukhov length—the vertical flux of heat, or by the difference in temperature between the sea surface and air temperature at 10 m height. The wind friction velocity, defined as

$$u* = (\tau/\rho_a)^{1/2} \tag{2}$$

where τ is the stress exerted on the ocean and ρ_a is the air's density, can describe most of atmosphere-ocean interactions (Massel 2007). It is strongly correlated with wind speed, sea state and stability in the lower part of the atmosphere. Summing up, it affects wave breaking formation.

SSA production can also be initiated by the rainfall. Impact of the raindrops on the ocean surface may cause the production of SSA particles, under certain conditions, like forming bubbles. Simultaneously rain is changing the temperature and salinity of sea water, removing surface film etc. (Yang et al. 1997). Other factors affecting SSA production and their later behavior described in the literature include some of interfacial phenomena, such as wind-wave interactions, gas exchange, whitecap formation and bubble dynamics.

To determine the aerosol emission from the sea surface, it has become necessary to term the source function (SF) of this phenomenon. However, the main problem are the differences in SF, dependent on research area; e.g. SF for aerosol flux on the Baltic Sea differs from the Northert Atlantic SF, according to Petelski and Piscozub (2006).

It was necessary to develop appropriate techniques to enable the measurement of the SSA emission and designation the source function. Several measurement methods are used to estimate the size-dependent SSA production from the sea surface shown clearly in Table 3, including examples of articles, where the given method was used and described.

The micrometeorological method is divided into several further methods described below:

- Eddy correlation/eddy covariance techniques, consists of simultaneous measurements of the vertical wind speed component and the component of interest *q* (*q* may be size-dependent SSA number concentration from which the flux can be determined). This method is effective on stationary ground. For example, this method was used by Weseley et al. (1981), Lee and Black (1994), Massman (1991), Buzorius et al. (1998).
- Eddy accumulation method is based on separate sampling the upwardly moving eddies and downwardly moving eddies with a volume flow rate, which is proportional to the magnitude of the vertical component of their velocity (Hicks and MacMillen 1984; Speer et al. 1985).
- Relaxed eddy accumulation method consists of collecting samples of upwardly and downwardly moving eddies at a constant rate. This method overcomes some difficulties of the eddy accumulation method. Proposed by Oncley et al. (1993), Nie et al. (1995).
- Gradient method, based on the difference between measured SSA particles concentration on two levels above the surface, used by, inter alia, Eriksson (1959), Gilette et al. (1972), Porch and Gilette (1977), Sievering (1981), Petelski (2003), Petelski et al. (2014).
- Inertial dissipation method, although not yet used to estimate the SSA emission, combine the assumptions of eddy correlation and gradient method and potentially also can be effective Proposed and described by Fairall and Larsen (1986), Fairall et al. (1990). Promising results of gas and aerosol measurements are described by Pryor et al. (2008), Sørensen and Larsen (2010).

Most recent advancements in field of SSA research area are shown below:

- Westervelt et al. (2012)—work showing the global emission source of sea spray and its impact on CCN concentration, with the SSA source function dependent on the local chlorophylle concentration.

Table 3 Methods for Determining Size-Dependent Sea Salt Aerosol Production Fluxes (Lewis and Schwartz 2004), and examples of publications where the given method was used

Method	Assumptions	Flux[a]	Size of particle[b]	Drop type[c]	Examples in literature
Steady state dry deposition	Size-dependent number concentration and dry deposition velocity	Eff	M	J	Moor and Mason (1954), Eriksson (1959), Blanchard (1963), Smith et al. (1993)
Whitecap	Oceanic whitecap ratio and size-dependent laboratory Whitecap production flux	Int	S	F	Cipriano et al. (1983), Woolf (1985), Monahan (1986), Stramska et al. (1990)
Concentration buildup	Size-dependent number concentration as a function of along-wind distance and height	Eff	S, M	F, J	Reid et al. (2001)
Bubble	Sizes and numbers of drops produced per bubble, size distribution of bubble concentration and bubble rise velocities	Int	S, M	F, J	Eriksson (1959), Blanchard (1963), Monahan (1993)
Micrometeorological	Size-dependent number concentration and size-dependent vertical fluxes at ~ 10 m	Eff	S	Film drops	Weseley et al. (1981), Rannik and Vesala (1999), Pattey et al. (1993), Duan et al. (1988), Petelski (2003)
Along-wind flux	Size-dependent along-wind flux near the sea surface and mean horizontal distance traveled	Int	L (S[d])	J, S (F)	Wu (1993)
Direct observations	Count from video	Int	L	S	Koga and Toba (1981)

(continued)

Table 3 (continued)

Method	Assumptions	Flux[a]	Size of particle[b]	Drop type[c]	Examples in literature
Vertical impaction	Size-dependent vertical flux near sea surface	Int	M, L	J, S	Bortkovskii (1987)
Statistical wet deposition	Size-dependent number concentration and mean time between precipitation events	Eff	S	F, J	Kritz and Rancher (1980), Cipriano et al. (1987)

[a]Effective (Eff) or Interfacial (Int) production flux
[b]Small (S) from ~0.1 to 1 μm range, Medium (M) from ~1 to ~25 μm or Large (L), bigger than 25 μm SSA particles
[c]Film (F), Jet (J) or Spume (S) drops type, according to applicability in chosen method
[d]In case of small SSA particles this method is identical as the statistical wet deposition method

- Grythe et al. (2014)—overview of 21 SSA existing functions, several of each are used in current climate models and source function proposed by authors.
- Ovadnevaite et al. (2014)—description of new sea spray source function, called OSSA—Oceanflux Sea Spray Aerosol, estimated on data from in situ-measurements collected on coastal station on the west coast of Ireland.
- Salter et al. (2014)—work based on laboratory experiment showing the connection between seawater temperature (from −1.3 to 30.1 °C) and aerosol production, in case of organic contamination free water, by using a plunging jet.
- Salter et al. (2015)—development of inorganic SSA function based on achievements described in Salter et al. (2014), parameterization for SSA production using a temperature-controlled chamber in laboratory.
- O'Dowd et al. (2015)—including the role of organic matter in surface layer and it's connection with the reflectance and cooling effect of bubbles bursting and they role in produce spray-droplets in the boundary layer.
- Wu et al. (2015)—parameterisation of sea spray generation function as wave-stante-dependent and wave-age-dependent phenomenon. The studies were applied to an atmosphere-wave coupled model.
- DeMott et al. (2016)—presenting the relationship between SSA emission and ice nucleating particles estimated by results from laboratory and measurements made over remote ocean regions. Showing association of SSA INP with phytoplancton blooms in laboratory stimulation.
- Witek et al. (2016)—showing impact of SSA parameterisation on the simulated aerosol optical depth (AOD) in the Southern Ocean region, by the simulations of SSA and measurements.

3 Our Methods and Results

The SSA particles were measured on the board of R/V Oceania, according to gradient method, based on Monin-Obukhov self-similarity theory (Petelski 2003). Area of sampling included Baltic Sea (southern part) in cruises in periods from October do May (Fig. 3.) and Norwegian and Greenland Seas in summer cruises AREX—Arctic Experiment (Fig. 4), both over the years. Measurements were collected by using Classical Aerosol Spectrometer (CSASP_100 HV) particle counter on five different levels: 8, 11, 14, 17 and 20 m a.s.l, (Zieliński 2004) under storm circumstances, when wind speed reaches over 10 m s^{-1}. Collecting time on every height takes 2 min, than the level is changed. During the measurements there can be no atmospheric fallout or fog, as it would affect results by causing wet deposition and changing the SSA particles density in the near-surface atmosphere. The vessel must be at the anchor, with prow toward the wind direction. Simultaneously the particles are measured by, e.g. Condensation Particle Counter (CPC)—TSI® model 3771 (diameter from 0.01 to 3 μm), Laser Aerosol Spectrometer (LAS)—TSI® 3340 (diameter from 0.09 to 7 μm), Open Path CO_2/H_2O Gas Analizer LI-7500A + WindMaster™ 3-Axis Ultrasonic Anemometer of Gill Instrumenter (Markuszewski et al. 2016) and five optical counter OPC-N2, attached at 8, 11, 14, 17 and 20 m above sea level. CSAPS_100 HV particle counter measure the diameters from 0.5 to 47 μm and OPC-N2 from 0.38 to 17 μm.

The results were well presented in publications: Petelski (2003)—description of gradient method and connection between SSA emission from sea surface and wind

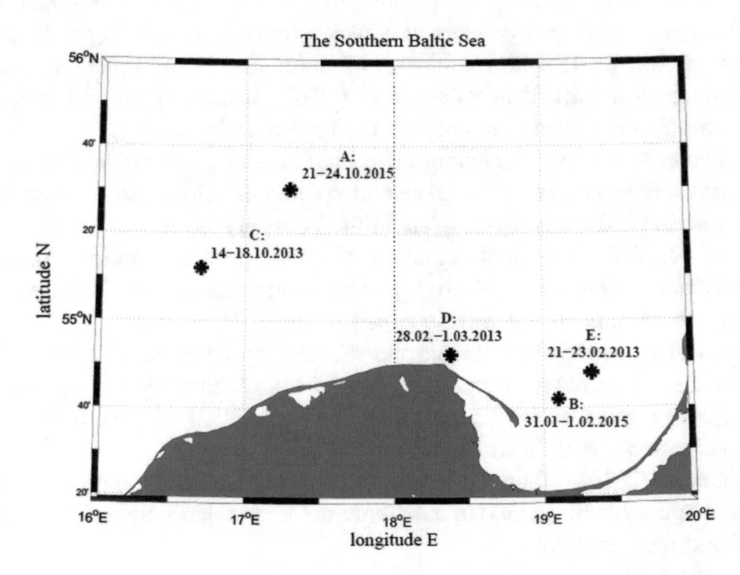

Fig. 3 Locations of the measurement stations (5 points in the Baltic Sea). A: the Southern Middle Bank, B and E: the Gdansk Deep region, C: the Slupsk Bank, D: off Cape Rozewie (Markuszewski et al. 2016)

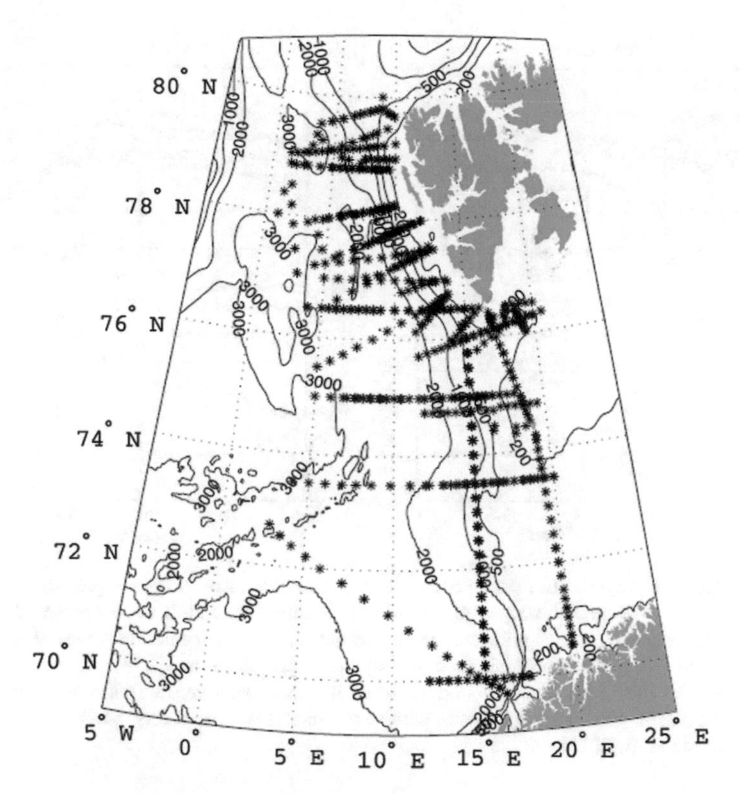

Fig. 4 Locations of hydrographic measurement stations during cruises of the R/V Oceania (AREX) in the Norwegian and Greenland Seas (with bathymetry data in meters) in summer periods in 1995–2003 (Petelski and Piskozub 2006)

speed, Petelski (2005)—describing the coarse aerosol particles by used the exponential function, in the near water atmospheric layer in Arctic regions, Petelski et al. (2005)—parameterization of the SSA emission using significant wave height and wind velocity, Petelski and Piskozub (2006)—coarse marine vertical aerosol flux source function calculated by used gradient method, for Arctic particles measurements, Petelski et al. (2014)—calculation of the Sea Salt Generation Function for the Baltic Sea depending on aerosol-size and wind speed, and Markuszewski et al. (2016)—comparision of measurements of SSA fluxes with data from the WAM model (Fig. 5).

Ability to use a research vessel gives us the opportunity to collect in situ measurements under appropriate weather conditions. The construction of R/V Oceania allows proper placement of the measurement apparatus that suits the specifics of our research. By using the gradient method, we have achieved good results in determining the North Atlantic source function. Moreover, as the only ones we have given the correct SF for the Baltic Sea—that is, the most comparable to the real one. Although it is an interesting research area, due to its size and low

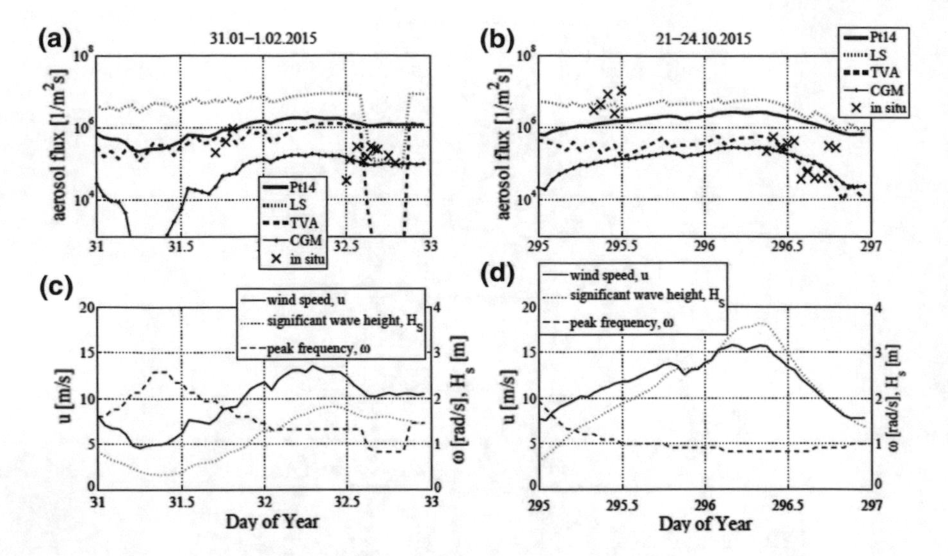

Fig. 5 **a** and **b** Time series of measured and modeled aerosol fluxes. Pt14 is a generation function dependent on wind speed only, LS (Limiting Steepness) and TVA (Threshold Vertical Acceleration) depends on significant wave height (H_S) and peak frequency (ω_p), CGM (Callaghan-Gong-Monahan) combines wind speed dependence and whitecap properties. Xs shows results of in situ measurements on board of the ship using vertical concentration gradient. **c** and **d** presents corresponding time series of modeled parameters used in calculations (Markuszewski et al. 2016)

salinity, the Baltic is usually discounted in scientific work devoted to the aerosol fluxes. Now we are focused on comparision of results from CSASP_100HV and OPC-N2 particle counters. Furthermore, parameterization and improvement of SSA source function is in progress, both for Baltic Sea and Northern Atlantic.

4 Conclusions

Marine aerosols are very important area of research, due to their vast impact on global climate and human's health as well as selected aspects of economy. For this reason scientists around the world are focusing on this particular topic, by carrying out tests and measurements in every available way, dependent on recent technological advancement and progress in technical sciences domain. Nonetheless, our knowledge of this complicated subject is still vestigial. We are aiming for a correct presentation of the aerosol source function, which can truly describe the reality around, against the algorithms describing climate change in laboratories and in model research. For this purpose it is becoming clearly essential to continue the in situ measurements by using particles counters, like OPC-N2, CSASP_100HV or others. By using a number of methods, described above, and by knowledge, experience and ideas exchange on the international ground, we will be able to continually enrich our knowledge in the field of SSA. Furthermore, it is important to find other factors that affect the aerosol emission from the sea surface, which are occurring in natural environment, but were not taken into account and more thoroughly studied before. All these actions, leading to clustering and parameterisation of marine aerosol source function, will ensure progress in this area in the future.

Acknowledgements Hereby, I would like to thank the Air-Sea Interaction team from the Institute of Oceanology Polish Academy of Sciences for facilitating my participation in sea cruises, collection and analyses of data.

References

Blanchard DC (1963) The electrification of the atmosphere by particles from bubbles in the sea. In: Sears M (ed) Progr Oceanogr, pp 73–202. Pergamon Press, New York

Bond TC, Streets DG, Yarber KF, Nelson SM, Woo JH, Klimont Z (2004) A technology-based global inventory of black and organic carbon emission from combustion. J Geophys Res D Atmos 109:D14203. https://doi.org/10.1029/2003JD003697

Bortkovskii RS (1987) Air-sea exchange of heat and moisture during storms, 194 pp. D. Reidel Publishing Company, Dordrecht

Boucher O, Randall D, Artaxo P, Bretherton C, Feingold G, Forster P, Kerminen V-M, Kondo Y, Liao H, Lohmann U, Rasch P, Satheesh SK, Sherwood S, Stevens B, Zhang XY (2013) Clouds and aerosols. In: Stocker TF, Qin D, Plattner G-K, Tignor M, Allen SK, Boschung J, Nauels A, Xia Y, Bex V, Midgley PM (eds) Climate Change 2013: the physical science basis. Contribution of Working Group I to the fifth assessment report of the intergovernmental panel on climate change. Cambridge University Press, Cambridge and New York

Buzorius G, Rannik Ü, Mäkelä JM, Vesala T, Kulmala M (1998) Vertical aerosol particle fluxes measured by eddy covariance technique using condensational particle counter. J Aerosol Sci 29:157–171

Cipriano RJ, Blanchard DC, Hogan AW, Lala GG (1983) On the production of Aitken nuclei from breaking waves and their role in the atmosphere. J Atmos Sci 40:469–479

Cipriano RJ, Monahan EC, Bowyer PA, Woolf DK (1987) Marine condensation nucleus generation inferred from whitecap simulation tank results. J Geophys Res 92(C6):6569–6576

Cofala J, Amann M, Klimont Z, Kupiainen K, Hoĺ-glund-Isaksson L (2007) Scenarios of global anthropogenic emission of air pollutants and methane until 2030. Atmos Environ 41:8486–8499

DeMott PJ, Hill TCJ, McCluskey CS, Prather KA, Collins DB, Sullivan RC, Ruppel MJ, Mason RH, … Franc GD (2016) Sea spray aerosol as a unique source of ice nucleating particles. Seinfeld JH (ed) PNAS, vol 113, 21: 5797–5803

Denterer F, Kinne S, Bond T, Boucher O, Cofala J, Generoso S, Ginoux P, Gong S, Hoelzemann JJ, Ito A, Marelli L, Penner JE, Putaud J-P, Textor C, Schulz M, van der Werf GR, Wilson J (2006) Emissions of primary aerosol and precursor gases in the year 2000 and 1750 prescribed data-sets for AeroCom. Atmos Chem Phys 6:4321–4344

Duan B, Fairall CW, Thomson DW (1988) Eddy correlation measurements of the dry deposition of particles in wintertime. J Appl Meteorol 27:642–652

Eriksson E (1959) The yearly circulation of chloride and sulfur in nature; meteorological, geochemical and pedological implications, Part 1. Tellus 11(4):375–403

Fairall CW, Larsen SE (1986) Inertial dissipation methods and turbulent fluxes at the ari-ocean interface. Bound-Layer Meteor 34:287–301

Fairall CW, Edson JB, Larsen SE, Mestayer PG (1990) Inertial-dissipation air-sea flux measurements: a prototype system using realtime spectral computations. J Atmos Oceanic Technol 7:425–453

Garbalewski C (2000) Fizyka Aerozolowej Aktywności Morza, Rozprawy i monografie 12/1999. Sopot, Poland, p 44

Gilette DA, Blifford IHJ, Fenster CR (1972) Measurements of aerosol size distributions and vertical fluxes of aerosols on land subject to wind erosion. J Appl Meteorol 11:977–987

Gong SL (2003) A parameterization of sea-salt aerosol source function for sub- and super-micron particles. Global Biogeochem Cycles 17(4):1097. https://doi.org/10.1029/2003GB002079

Grythe H, Ström J, Krejci R, Quinn P, Stohl A (2014) A review of sea-spray aerosol source functions using a large global set of sea salt aerosol concentration measurements. Atmos Chem Phys 14:1277–1297

Hicks BB, MacMillen RT (1984) A simulation of the eddy accumulation method for measuring pollutant fluxes. J Clim Appl Meteorol 23:637–643

Kettle AJ, Andreae MO, Amouroux D, Andreae TW, Bates TS, Berresheim H, Bingemer H, Boniforti R, Curran MAJ, DiTullio GR, Helas G, Jones GB, Keller MD, Kiene RP, Leck C, Levasseur M, Malin G, Maspero M, Matrai P, McTaggart AR, Mihalopoulos N, Nguyen BC, Novo A, Putaud JP, Rapsomanikis S, Roberts G, Schebeske G, Sharma S, Simo R, Staubes R, Turner S, Uher G (1999) A global database of sea surface dimethylsulfide (DMS) measurements and a procedure to predict sea surface DMS as a function of latitude, longitude, and month. Glob Biogeochem Cycles 13: 399e444

Koga M, Toba Y (1981) Droplet distribution and dispersion processes on breaking wind waves. Tohoku Geophys J 28(1):1–25

Kritz MA, Rancher J (1980) Circulation of Na, Cl, and Br in the tropical marine atmosphere. J Geophys Res 85(C13):1633–1639

Lee X, Black T (1994) Relating eddy correlation sensible heat flux to horizontal sensor separation in the unstable atmospheric surface layer. J Geophys Res 99(D9):18545–18553

Lewis ER, Schwartz SE (2004) Sea salt aerosol production. Mechanisms, methods, measurements and models. Geophys Monogr 152:1–4

Massel SR (2007) Ocean waves breaking and marine aerosol fluxes, pp 1–3. Springer, New York

Massman WJ (1991) The attenuation of concentration fluctuations in turbulent flow through a tube. J. Geophys Res 96(D8):15269–15273

Markuszewski P, Kosecki S, Petelski T (2016) Sea spray aerosol fluxes in the Baltic Sea region: comparision of the WAM model with measurements, Estuarine, Coastal and Shelf Science, ECSS-D-16-00044R1

Monahan EC (1986) The ocean as a source for atmospheric particles. In: Buat-Menard P (ed) The role of air-sea exchange in geochemical cycling. D. Reidel Publishing Company, Dordrecht, pp 129–163

Monahan EC (1993) Occurence and evolution of acoustically relevant sub-surface bubble plumes and their associated, remotely monitorable, surface Whitecaps. In: Kermann BR (ed) Natural Physical sources of underwater sound. Kluwer Academic Publishers, Dordrecht, pp 503–517

Moor DJ, Mason BJ (1954) The concentration, size distribution and production rate of large salt nuclei over the oceans. Q J Royal Meteorol Soc 80:583–590

Nie K, Kleindienst TE, Arnts RR, Sickles JEI (1995) The design and testing of a relaxed eddy accumulation system. J Geophys Res 100(D6):11415–11423

O'Dowd C, Ceburnis D, Ovadnevaite J, Białek J, Stengel DB, Zacharias M, Nitschke U, Connan S, Rinaldi M, Fuzzi S, Decesari S, Faccini MC, Marullo S, Santoleri R, Dell'Anno A, Corinaldesi C, Tangherlini M, Donovaro R (2015) Connecting marine productivity to sea-spray *via* nanoscale biological processes: Phytoplankton Dance or Death Disco? Sci Rep 5:14883. https://doi.org/10.1038/srep14883

Oncley SP, Delany AC, Horst TW, Tans PP (1993) Verification of flux measurement using relaxed eddy accumulation. Atmos. Environ. 27A(15):1426–2417

Ovadnevaite J, Manders A, de Leeuw G, Ceburnis D, Monahan C, Partanen A-I, Korhonen H, O'Dowd CD (2014) A sea spray aerosol flux parameterization encapsulating wave state. Atmos Chem Phys 14:1837–1852. https://doi.org/10.5194/asp-14-18837-2014

Pattey E, Desjardins RL, Rochette P (1993) Accuracy of the relaxed eddy-accumulation technique, evaluated using CO_2 flux measurements. Bound-Layer Meteorol 66:341–355

Petelski T (2003) Marine aerosol fluxes over open sea calculated from vertical concentration gradients. J Aerosol Sci 34:359–371

Petelski T (2005) Coarse aerosol concentration over the North Polar waters of the Atlantic. J Aerosol Sci Tech 39(8):695–700

Petelski T, Piskozub J (2006) Vertical coarse aerosol fluxes in the atmospheric surface layer over the North Polar Waters of the Atlantic. J Geophys Res. https://doi.org/10.1029/2005JC003295

Petelski T, Piskozub J, Paplińska-Swerpel B (2005) Sea spray emission from the surface of the open Baltic Sea. J Geophys Res 110:C10023. https://doi.org/10.1029/2004JC002800

Petelski T, Markuszewski P, Makuch P, Jankowski A, Rozwadowska A (2014) Studies of vertical coarse aerosol fluxes in the boundary layer over the Baltic Sea. Oceanologia 56(4):697–710

Porch WM, Gilette DA (1977) A comparision of aerosol and momentum mixing in dust storms using fast-response instruments. J Appl Meteorol 16:1273–1281

Pryor SC, Larsen SE, Sørensen LL, Barthelmie RJ (2008) Particle fluxes above forests: observations, methodological considerations and method comparisions. Environ Pollut 152:667–678

Rannik Ü, Vesala T (1999) Autoregressive filtering versus linear detrending in estimation of fluxes by the eddy covariance method. Bound-Layer Meteorol 91:259–280

Reid JS, Jonsson HH, Smith MH, Smirnov A (2001) Evolution of the vertical profile and flux of large sea-salt particles in a coastal zone. J Geophys Res 106(D11):12039–12053

Salter ME, Nilsson ED, Butcher A, Bilde M (2014) On the seawater temperature dependence of the sea spray aerosol generated by a continuous plunging jet. J Geophys Res Atmos 9052–9072. https://doi.org/10.1002/2013JD021376

Salter ME, Zieger P, Acosta Navarro JC, Grythe H, Kirkevåg A, Rosati B, Riipinen I, Nilsson ED (2015) An empirically derived inorganic sea spray source function incorporating sea surface temperature. Atmos Chem Phys 15:11047–11066

Sievering H (1981) Profile measurements of particle mass transfer at the air-water interface. Atmos Environ 15:123–129

Smith MJ, Park PM, Cinsterdine IE (1993) Marine aerosol concentrations and estimated fluxes over the sea. Quar J Royal Meteorol Soc 119:809–824

Sørensen LL, Larsen SE (2010) Atmosphere–surface fluxes CO_2 using spectral techniques. Bound-Layer Meteorol 136:59–81

Speer RE, Peterson KA, Ellestad TG, Durham JL (1985) Test of a prototype eddy accumulation for measuring atmospheric vertical fluxes of water vapor and particulate sulfate. J Geophys Res 90(D1):2119–2122

Stahler AH, Stahler AN (1992) Modern physical geography, 638 pp. Wiley, New York

Stramska M, Marks R, Monahan EC (1990) Bubble-mediated aerosol production as a consequence of wave breaking in super-saturated (hyperoxic) seawater. J Geophys Res 95(C10):18281–18288

Tsigaridis K, Koch D, Menon S (2013) Uncertainties and importance of sea spray composition on aerosol direct and indirect effects. J Geophys Res Atmos 118(1):220–235

van der Werf GR, Randerson JT, Collatz GJ, Giglio L, Kasibhatla PS, Arellano AF Jr, Olsen SC, Kasischke ES (2004) Continental-scale partitioning of fire emission during the 1997–2001 El Nino/La Nina Period. Science 303:73–76

Vignati E, Facchini MC, Rinaldi M, Scannell C, Ceburnis D, Sciare J, Kanakidou M, Myriokefalitakis S, Dentener F, O'Dowd CD (2010) Global scale emission and distribution of sea-spray aerosol: sea-salt and organic enrichment. Atmos Environ 44:670–677

Wesely ML, Cook DR, Williams RM (1981) Field measurement of small ozone fluxes to snow, wet bare soil, and lake water. Bound-Layer Meteorol 20:459–471

Westervelt DM, Moore RH, Nenes A, Adams PJ (2012) Effect of primary organic sea spray emission on cloud condensation nuclei concentrations. Atmos Chem Phys 12:89–101

Witek ML, Diner DJ, Garay MJ (2016) Satellite assessment of sea spray aerosol productivity: Southern Ocean case study. J Geophys Res Atmos. https://doi.org/10.1002/2015JD023726

Woolf DK (1985) Bubble and aerosol production at the ocean surface in Whitecaps and the marine atmosphere, Report No. 8. In: Monahan EC, Bowyer PA, Doyle DM, Higgins MR, Woolf DK (eds) University Collage, Galway

Wu J (1993) Production of spume drops by the wind tearing of wave crests: the search for quantification. J Geophys Res 98(C10):18221–18227

Wu L, Rutgersson A, Sahlée E, Larsén XG (2015) The impact of waves and sea spray on modelling storm track and development. In: Tellus A (ed) Dyn Meteorol Oceanogr 67 (1):27967. https://doi.org/10.3402/tellusa.v67.27967

Yang Z, Tang S, Wu J (1997) An experimental study of rain effects on fine structures of wind waves. J Phys Oceanogr 27:419–430

Zieliński T (2004) Studies of aerosol physical properties in coastal areas. Aerosol Sci Technol 38 (5):513–524

Discrete Element Method as the Numerical Tool for the Hydraulic Fracturing Modeling

Piotr Klejment, Natalia Foltyn, Alicja Kosmala and Wojciech Dębski

Abstract Hydraulic fracturing is an exploitation technique widely used in oil and gas industry. It is applied to unconventional reservoirs exploitation, where fracking must occur in order to enhance the flow of hydrocarbons. However, it is considered as a very invasive method. During a fracking process, huge amount of water with some additives is being pumped into a well and, as a result, creates a set of microcracks along the reservoir rock. That technique has been applied in renewable, geothermal energy, where fracking is simply releasing more thermal energy, previously captured in the rock. This article presents hydraulic fracturing phenomena described using Discrete Element Method (DEM). The basic assumption of DEM is that the rock can be represented as an assembly of particles joint into a rigid sample. Those particles are interacting with each other. Voids among particles simulate a pore system which can be filled with fracking fluid, which is numerically represented by much smaller particles. Following this microscopic point of view and its numerical representation by DEM method, primary results of numerical analysis of hydrofracturing phenomena are presented, using the ESyS-Particle Software. In particular, it is considered what is happening in distinct vicinity of the border between rock sample and fracking fluid, how cracks are creating and evolving by breaking bonds between particles. We conclude that numerical tools—as Discrete Element Method—have the potential to investigate mining technologies such as the hydraulic fracturing process.

Keywords Computer simulations · Numerical modeling · Hydraulic fracturing
Discrete element method · ESyS-Particle

P. Klejment (✉) · N. Foltyn · A. Kosmala · W. Dębski
Institute of Geophysics, Polish Academy of Sciences, Warsaw, Poland
e-mail: pklejment@igf.edu.pl

T. Zielinski et al. (eds.), *Interdisciplinary Approaches for Sustainable Development Goals*, GeoPlanet: Earth and Planetary Sciences,
https://doi.org/10.1007/978-3-319-71788-3_15

1 Hydraulic Fracturing—The Overview of the Method

1.1 The Introduction to the Hydraulic Fracturing Technique

Hydraulic fracturing is one of the key methods of enhancing extraction of unconventional oil and gas resources. These kinds of resources have lower permeability than conventional gas formations, therefore specific technologies, such as hydraulic fracturing, have to be used. Those resources can be extracted with the use of conventional methods, although fracturing vastly enhances flow of hydrocarbons (Resources to Reserves 2013).

The basic idea for hydraulic fracturing is that rock masses are fractured by a pressurized liquid. For this purpose, fracture fluid is being injected into a rock under the high pressure. Usually, fracture fluid consists of water with so-called proppant —additives as sand, ceramics or other thickening agent suspended throughout the bulk of the fluid. Fracture fluid is injected into a wellbore to create cracks in the deep rock formations (Fig. 1). As a result, hydrocarbons can flow more easily. After the hydraulic pressure is removed, the proppant becomes crucial. Its small grains hold the rock's fractures open.

Fracturing fluid is being pumped into a wellbore at a rate sufficient to increase pressure at the target depth. The aim of this procedure is that fracture gradient (pressure increase per unit of depth relative to density) of the rock has to be exceeded. When hydraulic pressure is higher than fracture gradient, the rock cracks. As a result, fracture fluid goes further and further along the fractures (Zoback 2007; Goodman et al. 2016).

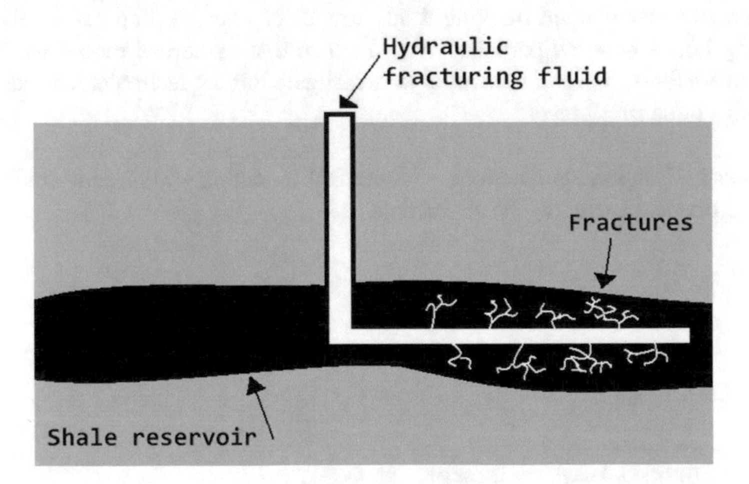

Fig. 1 Simplified scheme of the hydraulic fracturing process

During the process, migration of fracturing fluid from the fracture channel into the surrounding permeable rocks can occur (so called "leakoff'). In some cases it may cause serious loss of fracturing fluid. Consequently, the efficiency of the process can decrease. The location of one or more fractures along the length of the borehole is strictly controlled by various methods that create or seal holes in the side of the wellbore (Tan et al. 2017).

1.2 Environmental Impact of Hydraulic Fracturing

As every 'environmental-bonded' method of exploitation, hydraulic fracturing can affect the surroundings of the exploitation area. Opponents of the method argue that advantages are outweighed by the potential environmental impact.

Among the possible threats one can find **induced seismicity**, called microseismic events or microearthquakes (McGarr 2014). These microseismic events are often used to map the horizontal and vertical extent of the fracturing. The magnitudes of these events are usually too small to be detected at the surface. However, fluid disposal wells (which are often used in the US to dispose of polluted waste from several industries) have been responsible for earthquakes up to 5.6M (Holland 2011). That kind of quakes can cause damages in the structure of buildings. There are also concerns that quakes may damage underground gas, oil and water lines and wells that were not designed to withstand such pressure (King 2012; Ming 2017).

Hydraulic fracturing fluids include proppants and other substances, which may include toxic chemicals (Montgomery 2013). Consequently, there is a risk of **water contamination**. The danger is that the fracking fluid, if not protected properly, can penetrate not only the desired rock, but also the surroundings (www.epa.gov/hfstudy). While most of the additives are common and generally harmless, some chemicals are carcinogenic. Studies have shown that fracking fluid contains 55 carcinogenic chemicals which can not only get to the groundwater, but also pollute the air and the water during fracking fluid disposal (Vogel 2017).

Another problem connected with the hydraulic fracturing is the **water consumption**. Water usage by hydraulic fracturing can be a problem in areas that experience water shortage. Surface water may be contaminated through spillage or improperly built and maintained waste disposals. Further, ground water can be contaminated if fluid leakage occurs during fracking. Produced water, the water that returns to the surface after fracking, is managed by underground injection, municipal and commercial wastewater treatment, and reused in future wells (King 2012; Ming 2017).

Opponents of the hydraulic fracturing method emphasize the **consumption of large amounts of water** during even one fracking. Average hydraulic fracturing phenomena uses between 4500 and 13,200 m^3 of water per well, with large projects using up to 19,000 m^3 (King 2012; Ming 2017). Additional water is used when wells are refractured. An average well requires 11,000–30,000 m^3 of water over its lifetime (Jiang et al. 2014).

Development of the hydraulic fracturing method is an ongoing process. At the very beginning, hydraulic fracture efficiency was very low. Initially it was possible to extract only 2% of the gas stored in the deposit, now it reaches to 50% (Kasza 2011). The research over productivity enhancement is still under development. We propose to use computer simulations as an alternative way to study mining technologies and improve their efficiencies. Due to the fact that shales are usually brittle materials, we state that the Discrete Element Method (DEM)—designed for simulating rocks behavior—can be used for simulating some aspects of hydraulic fracturing process (Abe et al. 2002; Place et al. 2002). For instance, it fits ideally for analysis of cracks propagation.

2 Particle-Based Methods

Computer simulations are becoming important research tool in many scientific areas. So-called numerical modeling can be a useful and relatively cheap tool to investigate different phenomena without necessity of conducting difficult and expensive experiments, e.g. in laboratories. Nowadays, having a broader knowledge, one can model practically everything, from atoms, through DNA, ending on huge structures like geological faults. The use of numerical modeling in science is widely known, as well as in industry. The oil, gas, and mining industry is not an exception here—models of mechanical properties of different geomaterials and physical properties of rocks are very much appreciated due to high costs of real experiments. Additionally, those numerical models can predict materials behavior—can describe its reactions under the given conditions—which is really desirable in mining applications.

There are plenty of different numerical methods and new ones are being developed all the time. A variety of numerical modeling approaches have been applied to solving hydraulic fracturing phenomena. Some commonly used methods are based on the displacement discontinuity method to take advantage of the efficiency provided by a boundary element method (Jeffrey et al. 2017). Also the Galerkin's finite-element approach can be used for fluid-flow modeling and the rock mass deformation can be simulated using the elastic displacement discontinuity method (DDM) (Ghassemi 2017). The finite-element method has been applied to facturing problems but requires meshing of the entire domain and typically requires a fine mesh near the fracture tip (Jeffrey et al. 2017).

In comparison to other numerical modeling approaches, the Discrete Element Method describes better evolution of fractures in brittle material. DEMs have been widely used to simulate single fracture at the local scale or at the field scale of fractured rock masses that contain a huge number of fractures. The main advantages of DEMs are explicitly depicting the fracture geometry in relative details and considering the progressive degradation of material integrity during degradation process (Zhao 2017; Jing et al. 2013; Lisjak and Grasselli 2014). Due to DEM advantages, we applied this technique in our research.

2.1 Basic Idea Behind Discrete Element Method

The Discrete Element Method DEM, originally called Distinct Element Method is one of numerical methods for simulating materials consisting of a large number of particles (Cundall and Strack 1979). The basic idea behind DEM is to represent the material as an assemblage of discrete particles interacting with one another. With advances in computer power and numerical algorithms for nearest neighbor sorting, it has become possible to numerically simulate millions of particles on a single processor. At each time step, the calculations performed in DEM alternate between integrating equations of motion for each particle, and applying the force—displacement law at each contact. Linear particle motion is governed by The Newtonian equation:

$$\ddot{\vec{r}}(t) = \vec{f}(t)/M \tag{1}$$

where $\vec{r}(t)$ and M are position of the particle and the particle mass, respectively, $\vec{f}(t)$ is the total force acting on the particle.

The particle rotation is governed by the Euler's equations

$$\tau_x^b = I_{xx}\dot{\omega}_x^b - \omega_y^b\omega_z^b(I_{yy} - I_{zz}) \tag{2}$$

$$\tau_y^b = I_{yy}\dot{\omega}_y^b - \omega_z^b\omega_x^b(I_{zz} - I_{xx}) \tag{3}$$

$$\tau_z^b = I_{zz}\dot{\omega}_z^b - \omega_x^b\omega_y^b(I_{xx} - I_{yy}) \tag{4}$$

where τ_x^b, τ_y^b, τ_z^b are the components of total torque $\vec{\tau}^b$, ω_x^b, ω_y^b, ω_z^b are the components of angular velocities $\vec{\omega}^b$, and I_{xx}, I_{yy}, I_{zz} are the three principle moments of inertia. In the 3D case, $I = I_{xx} = I_{yy} = I_{zz}$ (Landau and Lifshitz 1969).

A DEM simulation is started by generating a model, which results in spatially orienting all particles and assigning an initial velocity (Fig. 2). The forces which act on each particle are computed from the initial data and the relevant physical laws and contact models. All forces are added up to find the total force acting on each particle. An integration method is employed to compute the change in the position and the velocity of each particle during a certain time step from Newton's laws of motion. Then, the new positions are used to compute the forces during the next step, and this loop is repeated until the simulation ends (O'Sullivan 2011).

Discrete Element Methods are relatively computationally expensive, which limits either the length of a simulation or the number of particles. Several DEM codes take advantage of parallel processing capabilities to scale up the number of particles or length of the simulation. One of them is an open source software ESyS-Particle. Due to its advantages this programme was used in presented research.

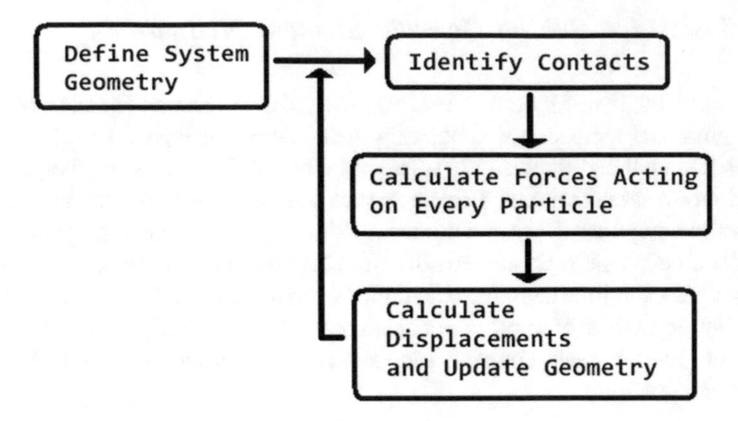

Fig. 2 The scheme of DEM simulation

2.2 *ESyS-Particle Software*

The ESyS-Particle is an open source software developed by the Earth Systems Science Computational Centre (ESSCC) at the University of Queensland (Abe et al. 2014). The engine of the software is written in C++ programming language and it has got implemented Message Passing Interface MPI. MPI divides performed simulation between multiple computers or multiple processor cores within the same computer. This provides the most efficient way of performing the simulation.

The main advantages of the ESyS-Particle were described (https://www.researchgate.net/post/What_are_the_different_advantages_of_three_DEM_open-source_software_Yade_ESyS-Particle_and_LIGGGHTS) by Dion Weatherley from University of Queensland, one of the lead developers: "*ESyS-Particle has two main advantages. The first is that its MPI parallel engine has been demonstrated to scale (weakly) to in excess 30,000 CPU cores and hence is definitely well suited for HPC DEM simulations. The second advantage is the BrittleBeamPrms particle-pair interactions that has been specifically designed for rock fracture and fragmentation simulations. This interactions correctly simulates Griffith crack propagation and, in particular, the formation of wing cracks. Credit to Dr. Yucang Wang* (Wang and Adhikary 2012) *for the mathematical formulation. Aside from the BrittleBeam interactions, ESyS-Particle provides a range of other interaction types*".

As it was mentioned above, to simulate elastic-brittle failure of rocks, sophisticated particle-pair interactions *BrittleBeamPrms* are used. This type of interactions incorporates both translational and rotational degrees of freedom. As illustrated in Fig. 3, two bonded particles may undergo normal and shear forces, as well as bending and twisting moments (Abe et al. 2014). Bonds designed to impart such forces and moments are known as cementatious bonds.

A Python wrapper API is another advantage of ESyS-Particle software. It provides flexibility in the design of numerical models, specification of modeling

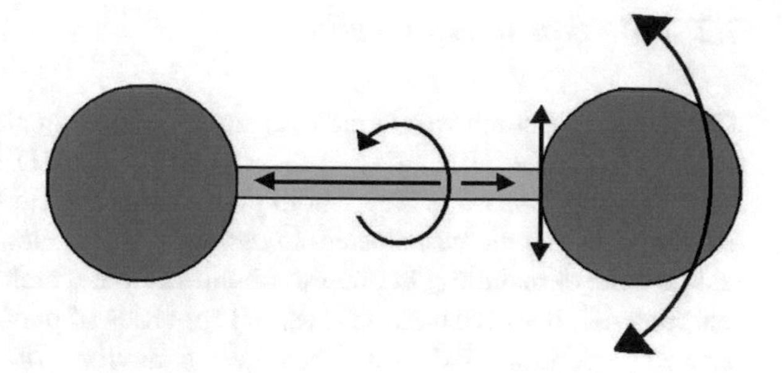

Fig. 3 Particle-pair interactions *BrittleBeamPrms*

parameters and contact logic, and analysis of simulation data. ESyS-Particle has been utilised to simulate earthquake nucleation, comminution in shear cells, silo-flow, rock fragmentation, and fault gouge evolution, to name a few applications. A wide range of different materials can be simulated, but the main goal of this research is to focus on geomechanics.

3 Attempts of the Discrete Element Method Application in the Area of Hydraulic Fracturing

3.1 Discrete Element Method in Mining Industry—Introduction

Discrete Element Method, whose basic idea was described in the previous paragraph, is one of the ways of obtaining a numerical model of structure or phenomena. Such simulations provide an alternative way of measuring the macroscopic properties of a synthetic material sample, as well as give an insight into its microscopic parameters.

DEM simulations can be useful in many areas connected with hydraulic fracturing itself (Yan et al. 2016), or in the drilling phase (which is adequate for the conventional and unconventional reservoirs). If the drilling is conducted properly, there is no possibility for fracture fluid to penetrate the surroundings. DEM simulations can demonstrate how the process of drilling will go, especially in the contact zone between bit and soil.

Application of DEM for the hydraulic fracturing itself is much more wide. One can get much information from just one model—cracks and fracking fluid propagation, pressure distribution in the rock during fracking, etc. (Wang and Adhikary 2012; Deng et al. 2014; Chong et al. 2017; Damjan and Cundall 2016). This data is very desirable since it cannot be obtained during real observations—fracking is performed in very deep formations, even few kilometers under the ground, and it is impossible to observe that phenomena in 'in situ' conditions (a camera can be used, but only visible results can be recorded, not physical).

3.2 Attempt of Application

The aim of this study was to recreate cracks spreading along shale reservoir (Fig. 1) during hydraulic fracturing process by using DEM implemented in the ESyS-Particle Software. The whole procedure consisted of several stages. First of all, model of sample was created. In case of discrete element based simulations, this task is quite demanding. In order to obtain sample which behavior correspond to the real material, iterative method is used. Properties of particles' parameters and bonds between particles should recreate behavior of material with specific macro-parameters, like Poisson's Ratio or Young's Modulus.

After this procedure, the model of the simulation was created and it represents fluid injected into a rock through a single hole in the main tube. It consists of two blocks of particles—upper block represented by smaller grains (radii from 0.1 to 0.3 mm) stands for fracturing fluid, meanwhile block of bigger particles (radii from 0.3 to 0.9 mm) represents brittle rock material (Fig. 4). Density of rock particles is 2.6 times bigger than density of fracturing fluid. Density of fracturing fluid is equal to 1 g/cm^3. It is a very simplified model of hydraulic fracturing where fracking fluid is represented as particles, which differ from those representing rock by density and size. Bonds between fluid's particles are broken easily after contact with almost three times denser rock mass. Then free fluid particles undergo frictional interactions and they can easily penetrate gaps between bigger particles of rock. Consequently, friction between small particles stands for fluid viscosity. Injection of these small particles into the block of bigger particles represents, in a simplified way, injection of fracking fluid during real hydraulic fracturing.

Different velocities were checked in order to find out how pressure changes during numerical simulation. One can distinguish three parts of the simulation. In the first one fluid is pumped into the rock formation which requires higher force because bonds between particles of the rock formation have to break. Shortly after that, pressure falls down up to zero. Bonds between particles in higher layer of the block were destroyed after first phase. Hence, fluid can flow easily into the newly opened space and, in this part of the simulation, nearly no force is needed to move the fluid. In the last, third part, all empty space in the upper part of the sample representing rock formation is fulfilled with fluid particles. There is no space for new ones so force is needed to pump another portion of liquid. As a result, in this phase pressure is rising.

Figure 5 represents results of simulations for different constant velocities. The pressure which pushes fluid particles into the rock was calculated with respect to the real time. One can notice three phases: (1) when fluid particles are getting closer to the rock and crashes into it; (2) after that pressure is dropping very fast to zero; (3) pressure is rising.

Different curves depict different velocities. Curvatures are very similar, except of one with the lowest pressure in the first phase, which is the slowest one—fluid velocity is 10 mm/s. In the second phase, major drop of pressure is observed for every velocity except of the slowest one again. The third phase shows more

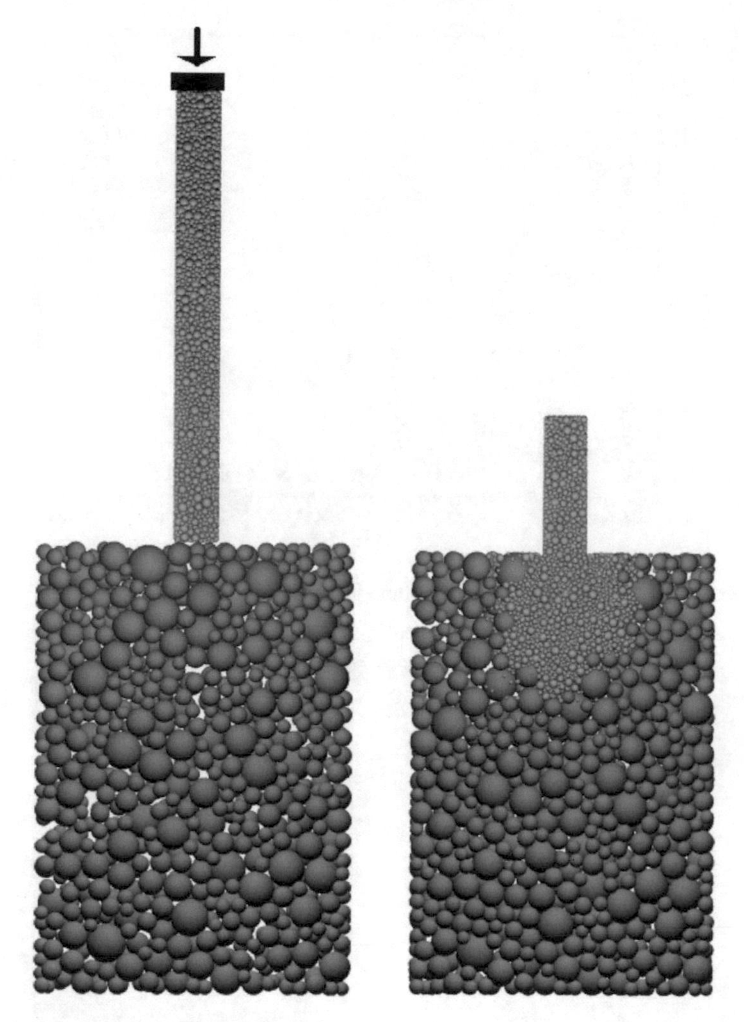

Fig. 4 DEM model of hydraulic fracturing process. Narrow block of particles imitates fracking fluid

differences between simulations, although they are similar in pressure rise. Again, simulation with the slowest velocity is not like the others.

Moreover, different way of moving the fracturing fluid was investigated under constant pressure. Block imitating the rock is surrounded by walls. This is three dimensional simulation but in reality it is more like cross section, because the thickness of this model is small in comparison with high and length. We applied such simplification to save simulation time. Below one can see forces acting on bottom wall in the Fig. 6. As it turned out, not the higher one (7.5 GPa) but middle pressure (5 GPa) is the most efficient in propagation of forces throughout the brittle sample.

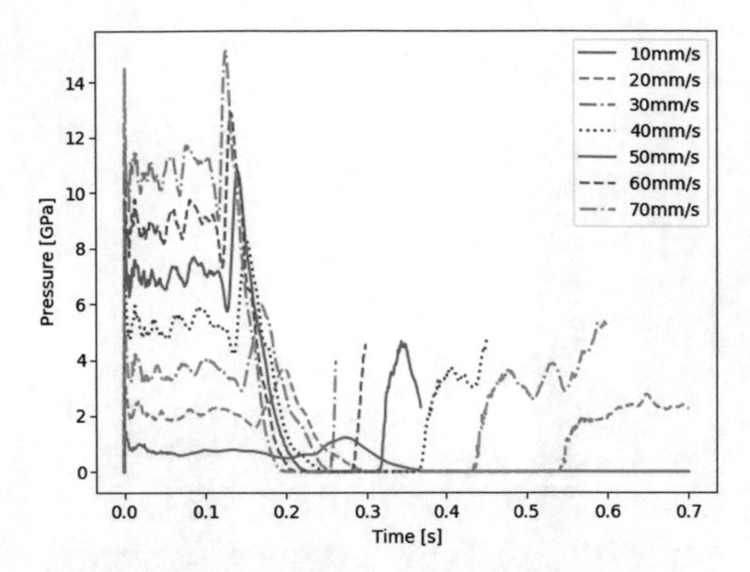

Fig. 5 Pressure acting on injected fracking fluid as a function of time

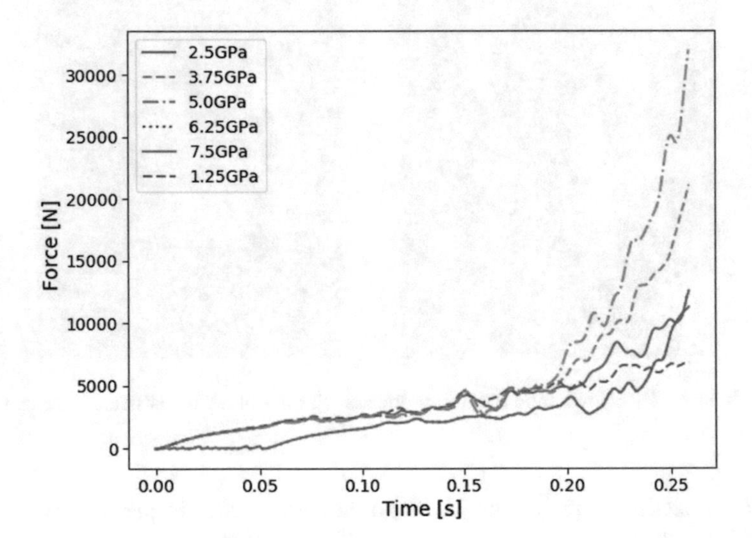

Fig. 6 Forces acting on the bottom wall

Figure 7 also represents constant forces of pressuring the fracturing fluid and velocity of this fluid (injection velocity) which were calculated with respect to time. For big forces of pressure one can notice some oscillations at the beginning of the simulation. It is caused by sort of "counter reaction" from the block with bigger particles and fracturing fluid particles or by some numerical instability. Fluid particles are moving but then they retreat.

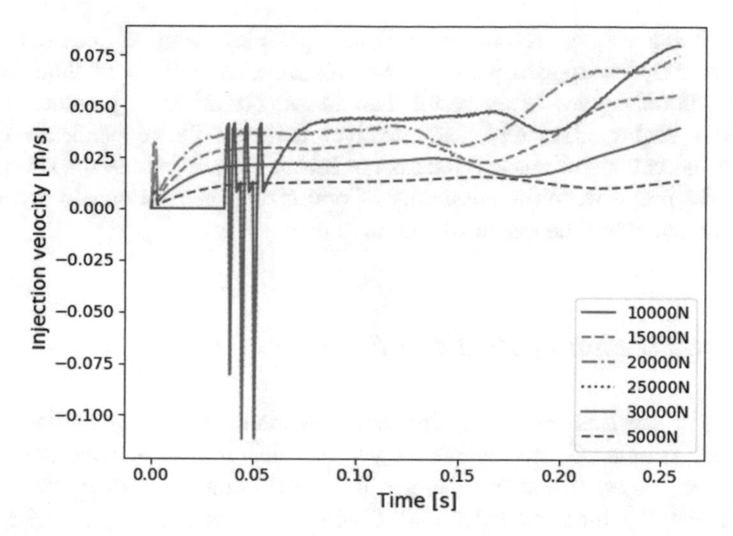

Fig. 7 Velocity of the fluid (injection velocity) calculated with respect to time

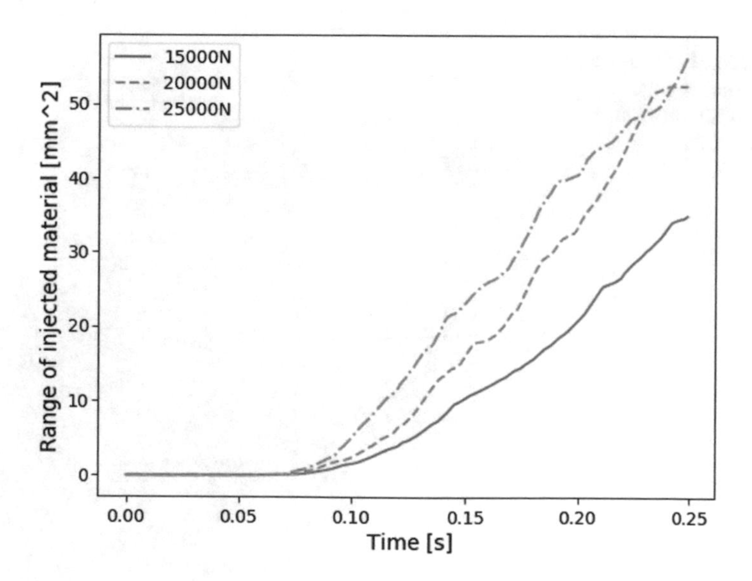

Fig. 8 The area in which fluid's particles appeared

Last but not least, Fig. 8 shows how far fluid particles can penetrate the rock. That gives the information how big cracks fracking phenomena can generate. For this purpose, the farthest positions of particles were found, and based on that information, the area of created cracks was calculated.

What is interesting, area in which fluid's particles occurred does not seem to depend directly on applied pressure. For instance, after 0.2 s of fluid injection, pressure 20000N gives better result than in case of 25000N pressure. Pressure 15000N is disproportionately less effective. It means that dependence between pressure and volume of cracked rock is not linear. This opens the space for further studies which may open the possibility of optimizing the relationship between the applied pressure and the extent of the fractures.

3.3 Investigation of Breaking Bonds

This section presents results of simulating hydraulic fracturing in the fully 3D cylindrical sample. Generally, model was divided into two main components: (1) Bigger, outer cylinder (height − h_{outer} = 20 mm and diameter − d_{outer} = 10 mm) and (2) Inner cylinder ended with cone (height − h_{inner} = 8 mm and diameter − $d_{inne}r$ = 5 mm) (Fig. 9).

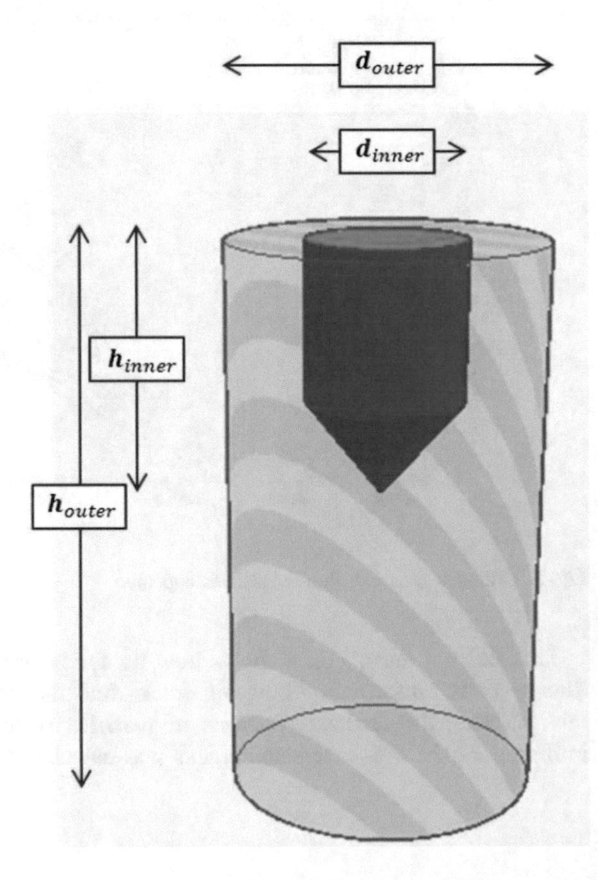

Fig. 9 Model of a simulation sample. Darker color represents inner cylinder with fracturing fluid

Fig. 10 Sample generated in ESyS-Particle Software. Dark particles represent hydraulic fracturing fluid, defined by smaller particles. Arrows show how force is applied during the simulation and circles represent loading wall; **a** view on the whole sample; **b** view inside the sample

Sample in ESyS-Particle Software was created based on model described above. Brighter particles represent rock formation, darker particles–fracking fluid (Fig. 10). Arrow indicates force applied on loading wall illustrated as a circle.

We conducted a series of different simulations in order to establish some basic properties of this numerical model. The two samples were created. One, presented in the Fig. 10 and the second one, exactly the same but with cylindrical mesh around it (visible in the Fig. 11). Particles represent rock have radius from 0.2 to 0.5 mm, meanwhile particles of fluid have 0.2 mm. Sample is simply limited by cylindrical mesh wall in order to avoid „barreling" and partially recreate in situ conditions. Number of bonds was taken under the investigation.

This value describes how many bonds exist in a model in particular numerical step. Those bonds are connecting particles with each other. We identify broken bond as a new fracture, so it is crucial to track all of them. Multiple sets of different parameters were investigated.

In the Fig. 12 one can observe a plot of a number of broken bonds with respect to numerical time steps, presenting six simulations with three different velocities of the wall, with and without cylinder mesh wall. Moving wall is illustrated in the

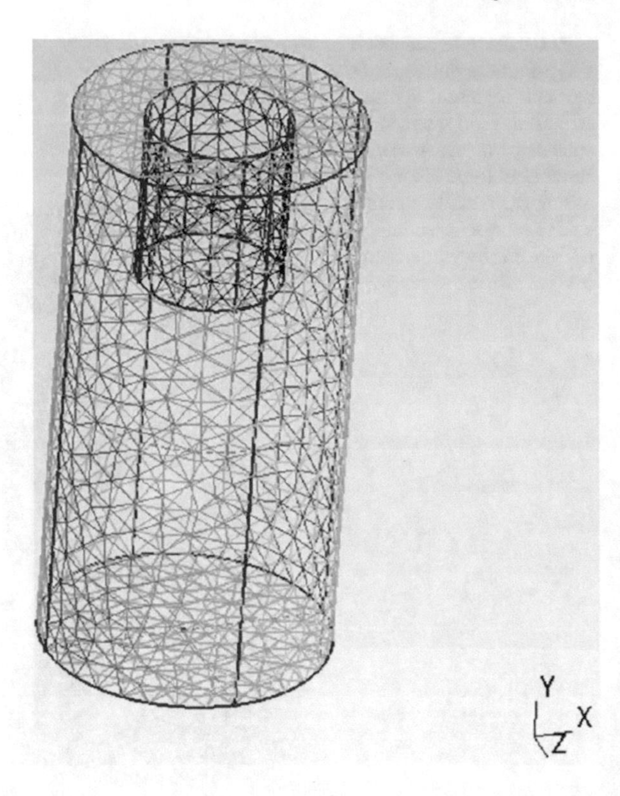

Fig. 11 Cylinder mesh wall

Fig. 10a and the Fig. 10b as a circle. The rest of the parameters remained unchanged. What is obvious, the bigger the velocity, the faster the bonds break. It can be observed, especially on a curve with biggest velocity, that the intensity of decrease of the number of bonds slows down. No further cracks are occurring, so network of cracks is created faster. When it comes to the cylinder mesh wall, it withstands cracks evolution. The smallest difference in percentage of broken bonds may be noticed for the lowest velocity. Also curvatures are similar. For other velocities, it is not observed.

Figure 13 illustrates three models without cylinder mesh wall, with the velocities 0.05 mm/s, 0.1 mm/s and 0.15 mm/s, respectively. For every model there are two snapshots at 500,000 and 1,300,000 numerical time step. One can observe disappearance in bonds between those snapshots. Due to „barreling" phenomenon, bonds tend to break horizontally rather than vertically.

When cylinder mesh wall is surrounding the sample, vertical kind of nature of bonds breakage can be observed more clearly. Of course, horizontally oriented bonds are appearing as well, but they are no longer in vast majority. Two snapshots of slices are presented above (Fig. 14). After comparing, differences between them are visible in the Fig. 15. Darker color indicates which bond has broken.

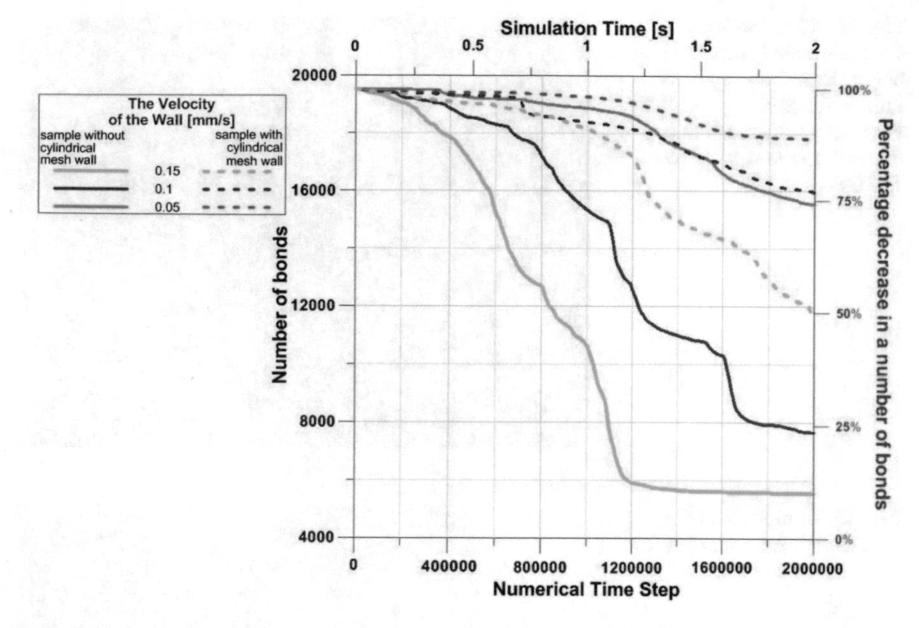

Fig. 12 Number of bonds as a function of numerical time step. Six different curves are presented, illustrating three different velocities of the wall, which "pushes" particles in inner cylinder into the sample, with and without cylinder mesh wall

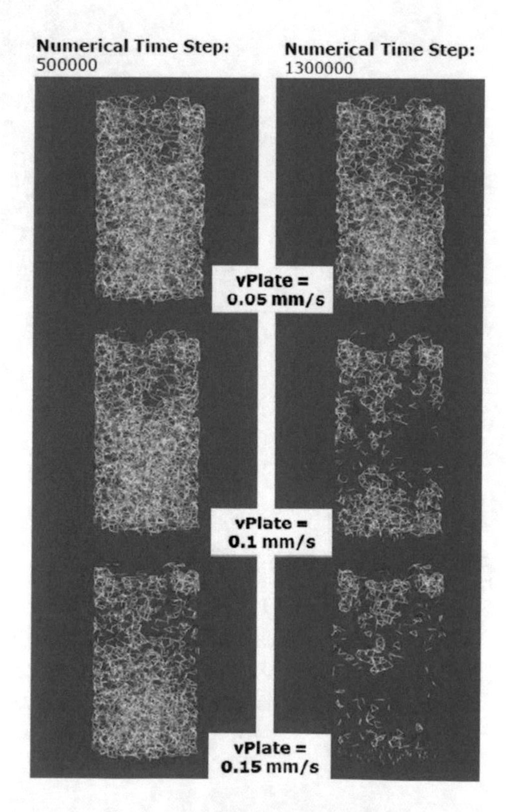

Fig. 13 Visualization of bonds. Snapshots represent samples in three velocities and in two numerical time step

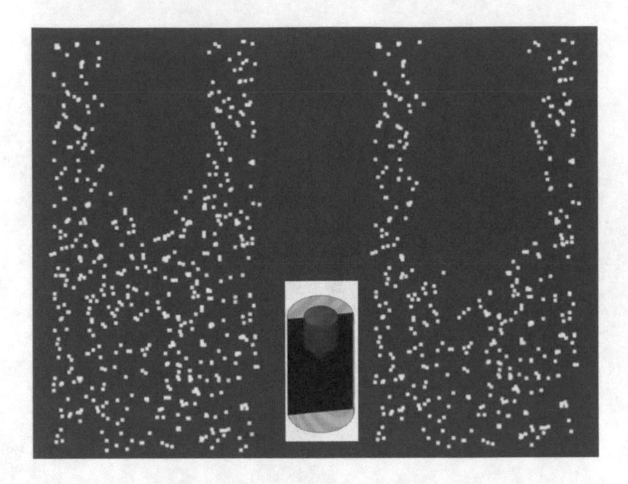

Fig. 14 Visualization of sample's cross section slice of bonds. Snapshots represent sample for one velocity = 0.15 mm/s in two numerical time steps, 0 and 130,000

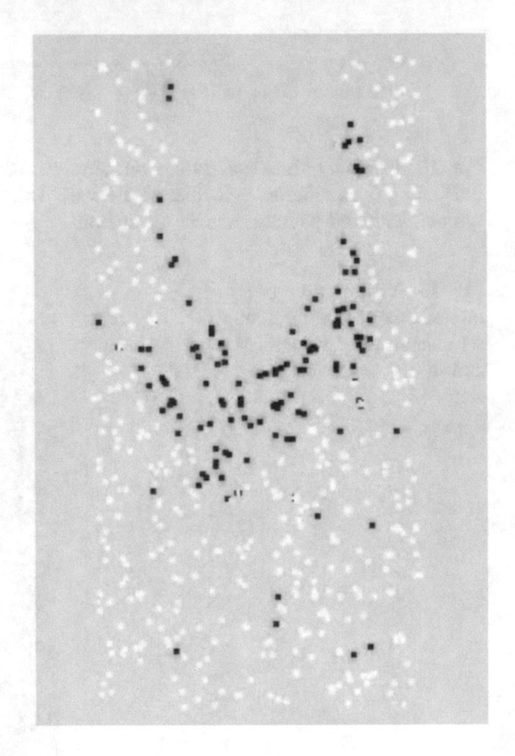

Fig. 15 Comparison of two snapshots presented in Fig. 15

4 Summary and Conclusions

We conducted a series of computer simulations trying to recreate, in a very simplified way, a well stimulation technique called hydraulic fracturing. The Discrete Element Method implemented in ESyS-Particle Software was used. The constructed model consists of great number of particles interacting with each other. Fracturing fluid was portrayed by much smaller particles in comparison with those building the rock. In Sect. 3.2 the results of fracking simulation are presented with rock portrayed as cuboids assembly of bigger particles, where in Sect. 3.3—as a cylindrical sample of bigger particles.

In Sect. 3.2 we investigated the behavior of brittle material into which fracking fluid was pumped with either constant velocity or constant pressure. Fluid, after breaking bonds within first layer of particles creating the rock, enters it freely. It was shown that the biggest pressure is needed to move the fluid into rock masses in the first phase, and in the second phase pressure drops to zero, in the last one it rises again. We also checked the relationship between the velocity of injecting fluid and the different values of constant pressure. Also, the area of penetration was calculated—area of created cracks.

In Sect. 3.3 we investigated the number of bonds between particles within cylindrical sample. Further studies demand implementing a method which would depict fluid kind of nature of fracturing fluid (mainly to take viscosity into account). Also parameters of the sample should be varied more, especially the range of particles radii. It has to be highlighted that it was an attempt and further studies must be carried out to fully understand this phenomenon.

Finally, it can be concluded that DEM simulations are a useful tool for studying hydraulic fracturing process. Hydraulic fracturing is extremely expensive technology. By using DEM simulations, it was possible to check how fracking fluid behaves inside rock formation under different constant pressures or velocities. Consequently, injection velocities were calculated for different values of pressure. It gives an insight into the sample which is possible to obtain only by using computer simulations. In conclusion we found the results very promising with a lot of possibilities for further research.

Acknowledgements We would like to thank professor Tomasz Danek and anonymous reviewer for their comments and valuable advice that have improved the article.

This research was supported by Young Scientist Grant Institute of Geophysics PAS no 6d/IGF PAN/2017 mł.

This research was carried out with the support of the Interdisciplinary Centre for Mathematical and Computational Modelling (ICM) University of Warsaw under grant No. GC70-6.

This paper was partially supported by research grant No. 2015/17/B/ST10/01946 of NCN, Poland.

References

Abe S, Boros V, Hancock W, Weatherley D (2014) ESyS-particle tutorial and user's guide version 2.3.1

Abe S et al (2002) Simulation of the influence of rate- and state-dependent friction on the macroscopic behavior of complex fault zones with the lattice solid model. Pure Appl Geophys 159(9):1967–1983

Assessment of the Potential Impacts of Hydraulic Fracturing for Oil and Gas on Drinking Water Resources Executive Summary, United States Environmental Protection Agency, External Review Draft | EPA/600/R-15/047a | June 2015 | www.epa.gov/hfstudy

Chong Z et al (2017) Numerical investigation of hydraulic fracturing in transversely isotropic shale reservoirs based on the discrete element method. J Nat Gas Sci Eng 46:398–420

Cundall PA, Strack ODL (1979) A discrete numerical model for granular assemblies. Geotechnique 29(1):47–65. https://doi.org/10.1680/geot.1979.29.1.47

Damjan B, Cundall P (2016) Application of distinct element method to simulation of hydraulic fracturing in naturally fractured reservoirs. Comput Geotech 71:283–294

Deng S et al (2014) Simulation of shale-proppant interaction in hydraulic fracturing by the discrete element method. Int J Rock Mech Min Sci 70:219–228

Ghassemi A (2017) Application of rock failure simulation in design optimization of the hydraulic fracturing. In: Shojaei AK, Shao J (eds) Porous rock fracture mechanics with application to hydraulic fracturing, drilling and structural engineering, pp 3–23. Woodhead Publishing Series in Civil and Structural Engineering, UK

Goodman PS et al (2016) Investigating the traffic-related environmental impacts of hydraulic-fracturing (fracking) operations. Environ Int 89–90:248–260

Holland A (2011) Examination of possibly induced seismicity from hydraulic fracturing in the Eola Field, Garvin County, Oklahoma. Oklahoma Geological Survey

https://www.researchgate.net/post/What_are_the_different_advantages_of_three_DEM_open-source_software_Yade_ESyS-Particle_and_LIGGGHTS. Accessed 28 Jul 2017

Jeffrey R, Zhang X, Chen Z (2017) Hydraulic fracture growth in naturally fractured rock. In: Shojaei AK, Shao J (2017) Porous rock fracture mechanics with application to hydraulic fracturing, drilling and structural engineering, pp 93–116. Woodhead Publishing Series in Civil and Structural Engineering, UK

Jiang M, Hendrickson CT, VanBriesen JM (2014) Life cycle water consumption and wastewater generation impacts of a Marcellus shale gas well. Environ Sci Technol 48(3):1911–1920. https://doi.org/10.1021/es4047654

Jing L et al (2013) Understanding coupled stress, flow and transport processes in fractured rocks. Geosyst Eng 16:2–25

Kasza P (2011) Zabiegi hydraulicznego szczelinowania w formacjach łupkowych. Nafta-Gaz XII:874–883

King GE (2012) Hydraulic fracturing 101: what every representative, environmentalist, regulator, reporter, investor, university researcher, neighbor and engineer should know about estimating frac risk and improving frac performance in unconventional gas and oil wells. SPE hydraulic fracturing technology conference, The Woodlands, 6–8 Feb 2012

Landau LD, Lifshitz EM (1969) Mechanics, volume 1 of a course of theoretical physics. Pergamon Press, Oxford

Lisjak A, Grasselli G (2014) A review of discrete modeling techniques for fracturing processes in discontinuous rock masses. J Rock Mech Geotech Eng 6(4):301–314

McGarr A (2014) Maximum magnitude earthquakes induced by fluid injection. J Geophys Res Solid Earth 119:1008–1019

Ming Q (2017) The impacts of fracking on the environment: A total environmental study paradigm. Sci Total Environ 580(15):953–957

Montgomery C (2013) Fracturing fluid components, effective and sustainable hydraulic fracturing. In: Jeffrey R (ed) InTech. https://doi.org/10.5772/56422

O'Sullivan C (2011) Particulate discrete element modeling: a geomechanics perspective [online], S.1: Spon Press/Taylor & Francis

Place D et al (2002) Simulation of the micro-physics of rocks using LSMearth. Pure Appl Geophys 159(9):1911–1932

Resources to Reserves (2013) Oil, gas and coal technologies for the energy markets of the future. International Energy Agency

Tan P et al (2017) Analysis of hydraulic fracturing initiation and vertical propagation behavior in laminated shale formation. Fuel 206(15):482–493

Vogel L (2017) Fracking tied to cancer-causing chemicals. CMAJ 16(189):E94–E95. https://doi.org/10.1503/cmaj.109-5358

Wang Y, Adhikary D (2012) Hydraulic fracture simulation based on coupled discrete element method and lattice boltzmann method. In: Proceedings World Geothermal Congress 2015. In: Li YG (ed) Imaging, modeling and assimilation in seismology. De Gruyter, Berlin

Yan Ch, Zheng H, Sun G, Ge X (2016) Combined finite-discrete element method for simulation of hydraulic fracturing. Rock Mech Rock Eng 49(4):1389–1410

Zhao Z (2017) Application of discrete element approach in fractured rock masses, pages 145–176; In: Shojaei A. K., Shao J., Porous rock fracture mechanics with application to hydraulic fracturing, drilling and structural engineering. Woodhead Publishing Series in Civil and Structural Engineering, United Kingdom 2017

Zoback MD (2007) Reservoir geomechanics. Cambridge University Press, New York

Comparing the Orientation of Geological Planes: Motivation for Introducing New Perspectives

Michał Michalak and Grzegorz Bytomski

Abstract This article discusses the problem of comparing the orientation of geological planes. We suggest two alternative approaches based on an analytical model using the coordinates of three non-collinear points to document a stratigraphic horizon. The methodological portion of this paper addresses selected topological conditions that should be satisfied such that the introduced tools are compatible with certain basic expectations and intuitions. The authors note that use of the suggested models must be preceded by adequate mathematical proofs. Because this research paper uses preliminary results, we briefly describe different methods of proving the desirable properties of the suggested functions and also present a counterexample associated with a function that is less computationally expensive. Selected basic calculations connected with measuring the angular distance between two investigated planes are also performed, and basic applications of the introduced tools are also described. The applications include assessment of the degree of nonconformities between stratigraphic horizons and the consequent detection of selected palaeo-environmental events of an erosive nature. We also discuss certain hydrogeological applications focused on detection of specific perturbations in unconfined aquifers. Because we offer two alternative approaches, we also present a comparison of the efficiency of the two models in different geological situations.

Keywords Comparing planes · Detecting nonconformities · Pseudometric Metric

---

M. Michalak (✉)
Faculty of Earth Sciences, University of Silesia, Sosnowiec, Poland
e-mail: mmichalak@us.edu.pl

G. Bytomski
Faculty of Mathematics, Physics and Chemistry, University of Silesia, Katowice, Poland
e-mail: grzbyt@wp.pl

T. Zielinski et al. (eds.), *Interdisciplinary Approaches for Sustainable Development Goals*, GeoPlanet: Earth and Planetary Sciences,
https://doi.org/10.1007/978-3-319-71788-3_16

1 Introduction and Motivation

The dip angle and dip direction are the two parameters used to sufficiently determine the orientation of geological planes. The dip angle specifies how steeply a plane dips, whereas the dip direction determines the direction towards which a plane dips. Among the many techniques and algorithms applied to this problem, a geological compass appears to be the most common tool used to obtain the orientation and is suitable for planes that appear in outcrops. However, certain other techniques can be used in obtaining these two angles. For instance, graphical methods that require an analysis of geological maps and estimated distances between contour lines (Groshong 2006). Certain analytical models require the coordinates of three non-collinear points to compute the dip angle and the dip direction (Groshong 2006; Vacher 1989). The set of three non-collinear points determines the equation of a plane and thus a three-dimensional normal (perpendicular) vector that represents the investigated plane in subsequent calculations (Vujicic and Sanderson 2008). In obtaining the orientation of a plane, the dip angle is related to the deviation from the horizontal position. Thus, we use a horizontal plane and an adequate vector. We denote the normal vector of the investigated plane by $\mathbf{v}$ and the vector of the horizontal plane by $\mathbf{u}$, and the dip angle is the angle between $\mathbf{v}$ and $\mathbf{u}$. This definition leads to a conclusion that the dip angle can be treated as a result of comparing the orientation between the examined plane and the horizontal plane. Nevertheless, the need often exists to measure the angular distance between two planes when neither of them is horizontal. For instance, to detect a fold in fieldwork, it is often sufficient for a geologist to find two planes that generally dip towards the opposite directions. This investigation can be performed using a geological compass and a spherical projection for estimation of the angle between two planes (Ragan 2009). The motivation for this work is that we wish to introduce two alternative tools for estimation of the difference between subsequent stratigraphic horizons and groundwater tables that can be proved in boreholes (Fig. 1). Because a horizon is a surface, it must be divided into triangular subareas whose shapes can be approximated as planar (Fig. 1). Once the triangulated irregular network (TIN) is generated, the adequate planes of triangular shape can be compared in terms of their orientation. We note that the suggested tools must be treated as distance tools. First, this approach means that if the planes are parallel, then the distance between them is equal to zero. This condition could be even stronger if we state that the distance between two planes x and y is zero if and only if $x = y$. In fact, we have described the beginning conditions that are stated in the definitions of the pseudometric and metric. The second expected property is that the angular distance between plane X and Y must be the same as the distance between Y and X. It must also be assured that the triangle inequality is satisfied, which means that the angular distance measured directly between two vectors $\mathbf{v}$ and $\mathbf{w}$ must always be smaller than the sum of the distances between $\mathbf{v}$ and $\mathbf{u}$ and between $\mathbf{u}$ and $\mathbf{v}$ for an arbitrary vector $\mathbf{u}$. Because the desirable properties are already discussed, the tools can be described more precisely. Obviously, the tools are mathematical functions, and the first

Fig. 1 Motivation for introducing the metric tools (rendered in POV-RAY)

example is simply the angle between two planes. In fact, this approach is often used by geologists, but this use is generally limited to measurements performed by a geological compass. Obviously, algorithms exist that allow us to perform the measurements numerically (Allmendinger et al. 2011). These algorithms use the dip angle and the dip direction as input data, which makes them incompatible with the procedure that requires coordinates for processing. As previously mentioned, our ultimate goal is to measure the angle using the coordinates of three non-collinear points that lie within a plane as input data. Nevertheless, to render the study more comprehensive, we should present proofs showing that the three abovementioned properties are satisfied, which also addresses the second function. We consider *sinx* or, in certain cases, *sintx* functions, where t is a parameter greater than or equal to one, and x is the angle between two vectors less than or equal to $\pi/2$. The main advantage behind the sine is that we can obtain a more sensitive tool for estimating the difference in orientation between the investigated planes. The significance of the suggested models can be noted especially in structural geology when no geophysical data are present. Groshong (2006) notes that distinguishing between conformable contacts and low-angle nonconformities can be crucial to the correct interpretation of a map. From a palaeo-environmental point of view, it is interesting to estimate the degree of nonconformities between two stratigraphic horizons. We can compute the arithmetic mean of the similarity between two horizons based on the partial results obtained through measurements of the discordance between triangular indivisible planes. If the nonconformity was estimated as relatively high, certain palaeo-environmental interpretations can be performed. From the environmental point of view, it can be valuable to detect hydrogeological perturbations because the groundwater table of an unconfined aquifer should be generally parallel to the impermeable layer, and if not, certain conclusions related to the discordance should be offered. This situation might indicate that unexpected layers are present between the groundwater table and the top of the impermeable layer. Once the functions are properly defined, we can create a program whose main goal is to transform the set of outputs into a continuous interval of colours indicating the degree of difference in orientation. For instance, if the two investigated planes are parallel, the program can cover the upper plane with blue, and if they are perpendicular, the colour of the upper plane is red. Another approach could treat the distance figures as attributes of one arbitrary horizon and display them as inseparable items accompanying the divided surface.

2 Methods and Algorithms

The technical aspect of this project uses programming techniques. Because the well-known computational geometry package "CGAL, Computational Geometry Algorithms Library" (2017) is available for C++ programs, C++ is used during implementation of the Delaunay triangulation and is responsible for generating the TIN. Certain proofs require visualizations, and these are performed in POV-RAY (Persistence of Vision Pty. Ltd. 2004).

From a computational point of view, the following procedure is used to compute the angle between two planes represented in the calculations by their normal vectors. The coordinates of the normal vectors of the examined planes can be obtained using the cross-product of the vectors. Assume that we have three non-collinear points $P_1 = (x_1, y_1, z_1)$, $P_2 = (x_2, y_2, z_2)$, and $P_3 = (x_3, y_3, z_3)$. Because the vectors $\overrightarrow{P_1P_2}, \overrightarrow{P_3P_1}$ lie in the plane, the vector $\overrightarrow{P_1P_2} x \overrightarrow{P_3P_1}$ is normal to the plane. To obtain the equation of the plane, it suffices to calculate a determinant (Vujicic and Sanderson 2008):

$$\begin{vmatrix} x_2 - x_1 & x_3 - x_1 & X \\ y_2 - y_1 & y_3 - y_1 & Y \\ z_2 - z_1 & z_3 - z_1 & Z \end{vmatrix} = 0 \tag{1}$$

Suppose that we have already calculated this determinant, and the two planes are given by following equations: (1) $aX + bY + cZ = 0$ and (2) $dX + eY + fZ = 0$. The normal vectors of these two planes are: $\mathbf{v} = [a, b, c]$ and $\mathbf{w} = [d, e, f]$. The angle between $\mathbf{v}$ and $\mathbf{w}$ is calculated as follows.

$$\arccos\frac{|v \cdot w|}{[\![v]\!][\![w]\!]} \tag{2}$$

where $|v \cdot w|$ is the absolute value of the dot product between vectors **v** and **w**, and is the product of lengths of vectors **v** and **w**. Because we expect the angle and its sine to describe selected topological properties, we introduce the definitions of the metric and pseudometric (Willard 2004).

Definition (*metric*):

If S is a set and $d : S \times S \rightarrow [0, \infty]$ is a function that satisfies:

$$d(S_1, S_2) = 0 \Leftrightarrow S_1 = S_2, \forall S_1, S_2 \in S \tag{3a}$$

$$d(S_1, S_2) = d(S_2, S_1), \forall S_1, S_2 \in S \tag{3b}$$

$$d(S_1, S_2) \leqslant d(S_1, S_3) + d(S_3, S_2), \forall S_1, S_2, S_3 \in S, \tag{3c}$$

then we say that d is a *metric* and a pair (S, d) is a *metric space*.

Definition (*pseudometric*):

If S is a set and $d : S \times S \rightarrow [0, \infty]$ is a function that satisfies:

$$d(S_1, S_1) = 0, \forall S_1 \in S \tag{4a}$$

$$d(S_1, S_2) = d(S_2, S_1), \forall S_1, S_2 \in S \tag{4b}$$

$$d(S_1, S_2) \leqslant d(S_1, S_3) + d(S_3, S_2), \forall S_1, S_2, S_3 \in S, \tag{4c}$$

then we say that d is a *pseudometric* and a pair (S, d) is a *pseudometric space*.

The choice between a pseudometric and metric does not appear to be of great importance from a computational point of view, but it is crucial for building a mathematical model that is a source of subsequent proofs. Two general approaches are used to prove the expected properties. The first approach uses the direct method of proof involving operations on vectors and their projections, whereas the second approach is related to the identification of angles as geodesics on a sphere. The identification makes the proofs easier to construct. Nevertheless, to make the study more comprehensive, we also present the direct method of proving the properties.

3 Results

Because the number $\frac{|v \cdot w|}{[\![v]\!][\![w]\!]}$ is simply the cosine of the angle x between two vectors **v** and **w**, one could consider the function $f(x) = 1 - cosx$ as a less computationally expensive equivalent of the suggested sine function. However, we also present a simple counterexample using the triangle inequality. Suppose we have two perpendicular planes S_1, S_2 and a plane S_3 as a bisecting plane between S_1. and S_2 Thus, the angle between S_1 and S_2 is $\pi/2$ and $\pi/4$ between S_1 and S_3 and $\pi/4$ between S_2 and S_3. Thus,

$$1 = f(\pi/2) > f(\pi/4) + f(\pi/4) \approx 0.58 \tag{5}$$

We present an example of calculating the angle between two planes P and R in which the sine is also calculated. We begin the calculations by supplying the coordinates of three points that generate both planes.

$$P_1 = (1, 2, 1) \tag{6a}$$

$$P_2 = (1, 1, 2) \tag{6b}$$

$$P_3 = (0, 2, 3) \tag{6c}$$

$$R_1 = (1, 3, -2) \tag{7a}$$

$$R_2 = (1, 0, -1) \tag{7b}$$

$$R_3 = (0, 1, -3) \tag{7c}$$

Subsequently, we can build the vectors that generate that plane:

$$\overrightarrow{P_1P_2} = [0, -1, 1] \tag{8a}$$

$$\overrightarrow{P_3P_1} = [-1, 0, 2] \tag{8b}$$

$$\overrightarrow{R_1R_2} = [0, -3, 1] \tag{9a}$$

$$\overrightarrow{R_3R_1} = [-1, -2, -1] \tag{9b}$$

To obtain the XYZ coefficients, we insert the vectors $\overrightarrow{P_1P_2}, \overrightarrow{P_1P_3}$ and $\overrightarrow{R_1R_2}, \overrightarrow{R_1R_3}$ into the matrices and calculate the determinants:

$$\begin{vmatrix} 0 & -1 & X \\ -1 & 0 & Y \\ 1 & 2 & Z \end{vmatrix} = 0 \tag{10}$$

and

$$\begin{vmatrix} 0 & -1 & X \\ -3 & -2 & Y \\ 1 & -1 & Z \end{vmatrix} = 0 \tag{11}$$

The equations of the planes are given:

$$-2x - y - z = 0 \tag{12}$$

and

$$5x - 3y - 3z = 0 \tag{13}$$

The normal vectors of the generated planes are given:

$$v = [-2, -1, -1] \tag{14}$$

and

$$w = [5, -3, -3] \tag{15}$$

Because the "z" coordinates of both vectors are negative, it indicates that the vectors are directed downwards. Because we only consider vectors directed upwards, all coordinates must be multiplied by −1. Thus, we obtain:

$$v_{\uparrow} = [2, 1, 1] \tag{16}$$

and

$$w_{\uparrow} = [-5, 3, 3] \tag{17}$$

The angle between these two planes is obtained from the formula:

$$\arccos \frac{|v_{\uparrow} \cdot w|}{[\![v_{\uparrow}]\!] [\![w]\!]} = \arccos \frac{4}{\sqrt{258}} \approx 75.58^{\circ}. \tag{18}$$

Once the angle computed, its sine is calculated. Thus, $sin(75.58^{\circ}) \approx 0.97$.

4 Discussion

In the presented example, we obtained the angle between planes generated by points. Because the interval [0, 1] is the set of outputs of the sine function considered on the interval $[0, \pi/2]$ and zero indicates total conformity (parallelism), the obtained result can be treated as a great nonconformity between the investigated planes. Actually, such situations seldom arise. Usually, the subsequent stratigraphic horizons show a similar dip. Obviously, computing the difference in orientation of similarly dipping planes using the simple angle metric appears pointless to visualize. The problem is that transforming the [0, 1] interval into a continuous scale of colours leads to covering the triangles with highly similar colours that are indistinguishable by the human eye. Thus, to perform a better visualization, users can use a function *sintu* for $t \geq 1$, and $u \leq \pi/2$. Obviously, t must be selected such that the function *sintu* considered on the interval $[0, u]$ is monotonous. The discussed models can have great significance in structural geology, where the need exists for estimation of the degree of nonconformities between subsequent surfaces proved in boreholes. It appears that such tools have not yet been defined. However, their significance appears to be crucial to the subsequent palaeo-environmental interpretations associated with marine transgressions or other events of an erosive nature. Because the triangulated irregular network can be generated for beds and for groundwater tables, hydrogeology could also take advantage of the suggested models. The investigation of potential nonconformities between the groundwater table and impermeable layer lying at a lower position appears to be of great significance.

5 Conclusion

Comparing the orientation of geological planes based on the coordinates of three non-collinear points appears to be a challenging task for the conventional method of computing the angle between two planes. The traditional method requires a time-consuming spherical projection and does not compute the angle between two investigated planes based on the coordinates of points lying within these planes. Thus, our approach differs from the conventional method that requires the dip angle and the dip direction as input data. The objective of this study was to describe tools that could be used in visualization of the difference in orientation of geological surfaces. These visualizations can be understood either as a display of numbers as attributes on one arbitrary plane or a display that covers the plane with colours related to the degree of the nonconformity that was proved between the two planes. However, introduction of certain tools must be preceded by information on their correctness and efficiency. The correctness can be associated with presentation of certain mathematical proofs, and more than one method of proving the required properties exists. Efficiency cannot be uniquely defined because it depends on the geological structure of the investigated area. It appears that the most natural metric, i.e., the angle between two planes, should be modified in most cases to supply better visualizations of the measurements. The environmental applications are focused on detection of nonconformities that can be associated with erosive events in the geological past. These tools might also be used in detection of perturbations in unconfined aquifers. The project described above includes tasks that require integration of many scientific fields, and we hope that our interdisciplinary project opens new unexpected frontiers.

Acknowledgements The authors are grateful to Prof. Przemysław Koprowski and Dr. Paweł Gładki for helpful remarks on different methods of proving the desirable properties. The authors also appreciate the comments from the anonymous reviewers who helped to improve this paper.

References

Allmendinger R, Cardozo N, Fisher D (2011) Structural geology algorithms: vectors and tensors. Cambridge University Press

CGAL, Computational Geometry Algorithms Library (2017) http://www.cgal.org/. Accessed 1 Sept 2017

Groshong RH (2006) 3-D Structural geology: a practical guide to quantitative surface and subsurface map interpretation. Springer-Verlag, Berlin Heidelberg

Persistence of Vision Pty. Ltd. (2004) http://www.povray.org/download/. Accessed 1 Sept 2017

Ragan DM (2009) Structural geology: an introduction to geometrical techniques. Cambridge University Press

Vacher HL (1989) The three-point problem in the context of elementary vector analysis. J Geol Educ 37(4):280–287. https://doi.org/10.5408/0022-1368-37.4.280

Vujicic M, Sanderson J (eds) (2008) Linear algebra thoroughly explained. Springer-Verlag, Berlin Heidelberg

Willard S (2004) General topology. Mineola, New York

Printed in the United States
By Bookmasters